D0882866

What Science Knows

WHAT SCIENCE KNOWS
KNOWS
And How It Knows It

James Franklin

Encounter Books New York · London

Copyright © 2009 by James Franklin

All rights reserved. No part of this publication may be reproduced,
stored in a retrieval system, or transmitted, in any form or by
any means, electronic, mechanical, photocopying, recording,
or otherwise, without the prior written permission of
Encounter Books, 900 Broadway, Suite 601,
New York, New York, 10003.

First American edition published in 2009 by Encounter Books,
an activity of Encounter for Culture and Education, Inc.,
a nonprofit, tax exempt corporation.
Encounter Books website address: www.encounterbooks.com

Manufactured in the United States and printed on
acid-free paper. The paper used in this publication meets
the minimum requirements of ANSI/NISO Z39.48-1992
(R 1997) (*Permanence of Paper*).

FIRST AMERICAN EDITION

LIBRARY OF CONGRESS CATALOGING-IN-PUBLICATION DATA
Franklin, James, 1953-
What science knows : and how it knows it / by James Franklin.
p. cm.
Includes bibliographical references and index.
ISBN-13: 978-1-59403-207-3 (hardcover : alk. paper)
ISBN-10: 1-59403-207-6 (hardcover : alk. paper)
1. Science—Philosophy. I. Title.
Q175.F7855 2009
501—dc22
2009006496

10 9 8 7 6 5 4 3 2

CONTENTS

PREFACE

Any time is a good time to contemplate the advances of science. But the ideal occasion is a visit to the dentist. Not only is the distraction welcome, but the intrusion of drill or laser into the mouth—so close to where he feels one's real self is located—prompts reflection on how much worse things could be. Or how much worse they actually were, before science worked its magic.

Take the case of Charles Whitworth (1752–1825). The state of Earl Whitworth's teeth as of 1825 is known exactly because he was buried in a triple-shelled lead coffin, excavated in the 1990s. He had been British ambassador to Napoleon and Lord Lieutenant of Ireland and could afford the highest standard of dental care his age could provide. That care, unfortunately, included the provision of tooth powders and tinctures of the sort advertised to "eradicate the scurvy and tartar from the gums; make the teeth, however yellow, beautifully white…." The reason teeth came up beautifully white was that the products contained abrasive materials made of shells, corals, and ground pebbles, along with tartaric acid—and Whitworth's teeth show the effects. On the front of the right upper incisors, where a right-handed man would naturally brush hardest, the enamel is completely missing. Having exposed dentine is a very painful condition, especially when trying to eat or drink anything cold or hot.[1]

A normal citizen of any functioning country today is the beneficiary of dental knowledge that Whitworth would have given his eye teeth for, what was left of them. We accept these benefits, and if we choose to, we can understand the science behind how they work.

Victims of land mines and napalm, it is true, are entitled to their vote that science sometimes has serious ill effects, and it is possible that a biology laboratory will yet come up with a microorganism that eats us all. It is the nature of science that it delivers power without responsibility. It delivers power because it delivers knowledge.

But scientific knowledge has many enemies. They resent not just the uses of science, but its aims, methods, and discoveries. The Internet and World Wide Web (provided by science, of course) are flooded with complaints and suspicion about science. Scientific theories, it is variously alleged, are socially constructed, determined by vested interests, underdetermined by data, dependent on the observer, logically impossible to confirm, always falsified in the long run, Western, godless, linear, patriarchal, reductive, and so on—all with the implication that scientific theories are not to be believed.

Science has its defenders too, many of them excellent at such individual tasks as refuting postmodernist attacks and defending particular aspects of science and the philosophy of science. What these defenders have not done is provide a simple and straightforward introduction to "why science is rational."

This book attempts that task. With a mixture of considerations about the logic of science and illustrations from real science, it explains from the ground up how science has established conclusions that are worthy of belief—absolutely certain conclusions in the case of the mathematical sciences, very highly probable ones in empirical science.

At the core of the defense of scientific rationality lies the objectivity of logical relations. This applies both to the deductive relations of mathematics and the probabilistic or nondeductive relations of empirical science. The reason we can prove—and hence believe with certainty—that the square of any even number is even is that deductive relations exist between truths about numbers and those about their squares. The reason we can rely on what the dentist tells us about our fate over the next hour—though with less than 100 percent certainty—is that there are logical relations between evidence and hypothesis. The evidence lies in the clinical trials that the dentist's materials, equipment, and procedures have undergone; the results of those trials bear on our case for logical reasons.

We begin, then, by explaining how an objective view of the relation of evidence to conclusion solves the classical problem of induction, which asks how we can know (with high probability) that all ravens are black when we have observed only finitely many black ravens and it is logically possible that the next raven we observe should be white. If that problem cannot be solved, there is not much hope for defending the rationality of the more esoteric reaches of science.

Basic though it is, the rationality of induction is sufficient for defending science against the broad-brush irrationalist attacks on it—by twentieth-century philosophy of science and, later, sociologists of science and postmodernists. An excursus on their objections reveals their logical mistakes and gross misunderstandings about the logic of evidence and conclusion.

We then survey some typical examples of knowledge in the natural, cognitive, and social sciences, to give some sense of the variety of methods used in real science. There is more than usual attention paid to the mathematical sciences, not only because they long ago found what is the gold standard of knowledge, mathematical proof, but also because the computer revolution has extended the reach of mathematical methods through most of science.

After a brief glance at how science as actually realized in people and institutions supports the discovery of scientific truths (or occasionally does not), we conclude with a view of the limits of science. Some of those limits are imposed by the problems of observing the very big, very small and very old, and understanding the very complex. In particular, the controversies about evolution and global warming arise from the inherent difficulty of understanding the complex systems involved. But beyond that there are more principled limitations to science, namely the essentially non-scientific character of some of the topics on which we most desire and need knowledge: consciousness and ethics.

Science has taught us not only what to think but how to think. Let us learn how it did it.

CHAPTER 1

Evidence

Some science is about hidden worlds, either smaller than the microscopic or more distant than our galaxy. Some involves such esoteric concepts as curved space-time or infinite dimensional Hilbert spaces. But the true lover of science will revel first in the many low-level empirical generalizations that summarize and give shape to our long perceptual experience: "All ravens are black"; "Last night's stars form much the same patterns as tonight's"; "Banana peels are normally slippery"; "Virgins don't have children"; "It's in spring that the wheat comes up"; "Lithium is effective against bipolar disorder"; "Tossed coins come up heads about half the time."

All Ravens Are Black: How Do They Know That?
An enormous amount of careful human observation has gone into recognizing and establishing these fundamental facts of science, which provide the foundations on which the edifice of more theoretical science rests. If we are to understand the rationality of science, we need to grasp first how we know those straightforward kinds of truths.

Logically speaking, there are three potential problems with establishing such simple generalizations:

- Do we have our classifications and concepts straight, so that we know definitely what potential new instances are or are not ravens and black?
- Have we established how we know by perception a single instance, "This raven is black"?
- How do we manage to make the leap of "inductive" inference from "all the (finite number of) observed ravens have been black" to "all ravens (including unobserved and future ones) are black"? Are we justified in doing so, and, if so, with what degree of confidence?

Those are all good questions. The first two, dealing with the scientific and logical underpinnings of our commonsense knowledge, will be considered in later chapters. The last one, the "problem of induction," has rightly been regarded as the classic problem that must be solved first in order to understand why science is rational.

To be in a position to solve it, one must first understand why seeing a black raven is evidence for the proposition that all ravens are black. We need to start at the beginning with the notion of evidence for beliefs.

EVIDENCE PRAISED

Don't you believe in flying saucers, they ask me? Don't you believe in telepathy?—in ancient astronauts?—in the Bermuda Triangle?—in life after death?

No, I reply. No, no, no, no, and again no.

One person recently, goaded into desperation by the litany of unrelieved negation, burst out, "Don't you believe in anything?"

"Yes," I said. "I believe in evidence. I believe in observation, measurement, and reasoning, confirmed by independent observers. I'll believe anything, no matter how wild and ridiculous, if there is evidence for it. The wilder and more ridiculous something is, however, the firmer and more solid the evidence will have to be."

Isaac Asimov, *The Roving Mind* (Amherst: Prometheus Books, 1997), 43.

It is wrong always, everywhere, and for anyone, to believe anything upon insufficient evidence.

William K. Clifford, "The Ethics of Belief" in *Lectures and Essays,* ed. Leslie Stephen and Sir Frederick Pollock (London: Macmillan and Co., 1879).

The Objective Bayesian View of Evidence

Objective Bayesianism, or logical probabilism, holds that the relation of uncertain evidence to conclusion is one of pure logic.[1] That the Big Bang theory is well-supported by present evidence, that a defendant's guilt has been proved beyond reasonable doubt, that there is good reason to believe well-confirmed conjectures in pure mathematics such as the Riemann hypothesis, are objective matters of the same nature as the deducibility of the Pythagorean theorem from Euclid's axioms.

Objective Bayesianism contrasts with *deductivism* (the thesis that all logic is deductive). Objective Bayesians argue for the existence of a non-deductive logic—a *logical* probability, or partial implication, between a body of evidence and a hypothesis that it supports or confirms but does not deductively imply.

The "probability" spoken of here is distinct from the probability involved in stochastic processes like throwing dice or tossing coins. More will be said on the kinds of probability in chapter 10.

Let us lay out the main arguments for the existence of non-deductive logic. Some of them are:

- The heated arguments among jurors, scientists, and others about whether particular bodies of evidence do render particular conclusions highly credible is a reason to believe that high credibility on evidence is a genuine relation, unless some better account of it can be given.

■ The inference scheme variously called the "proportional syllogism" or "statistical syllogism" or "direct inference"[2] certainly looks like the standard argument of deductive logic called the syllogism. The syllogism is the classic argument form of which an example is:

All men are mortal.
Socrates is a man.
So, Socrates is mortal.

The premises entail that the conclusion is true: it is impossible for the premises (All men are mortal, and Socrates is a man) to be true and the conclusion (Socrates is mortal) to be false. The "statistical syllogism" is an argument such as:

99.9 percent of men are mortal.
Socrates is a man.
So, Socrates is mortal.

In this instance, it is *possible*, but not likely, for the premises to be true and the conclusion false. What is so special about 100 percent, as opposed to 99.9 percent, that could make the first argument part of logic but the second not? A stipulation that the word "logic" should only apply to cases where it is impossible for the premises to be true and the conclusion false would simply evade the issue, which is whether propositions *can* support one another in ways that fall short of strict entailment. Evasions based on what would happen if extra premises (for example, "Socrates is divinely favored") were added to one or both of these inferences are no better, since the question, like any concerning inference, deals with the relation of the *given* premises to the conclusion, not the relation of some other set of premises to the conclusion.

- The simplest principle of logical probability, called by Polya "the fundamental inductive pattern"[3] (and the main content of the celebrated Bayes' theorem that gives Bayesianism its name), is:

q is a (non-trivial) consequence of hypothesis p.
q is found to be true.
So, p is more likely to be true than before.

It is hard to begin reasoning about the world without a commitment to that principle. Imagine a tribe that did not believe in it, and thought instead that agreement between theory and observation was a reason for *dis*believing the theory. Its members guess there are bison in the river field and go there to hunt them. They find none. So they conclude they will probably find bison there tomorrow and the next day and they go there day after day with high hopes. You will need to imagine that tribe because you will not be meeting them. They are extinct.

- The evaluation of conjectures in pure mathematics uses the usual non-deductive inference schemes such as the confirmation of theories by their consequences. There are inductive arguments used in experimental mathematics, such as conjectures that arise from observations of the digits of π. (More in chapter 10.) Because mathematics is true in all possible worlds, the rationality of these inferences must be matters of logic.
- The concept of "inference to the best explanation" (more later in this chapter) is widely regarded as necessary to make the most basic inferences about the existence of the external world, the existence of laws of nature, of atoms, and so on. Such inferences are obviously non-deductive,

and if they depended on any contingent facts, such as the existence of laws of nature, they could not perform the tasks assigned to them.

SOME FORMULAS (OPTIONAL)

There is no need for jurors evaluating evidence in a trial to know formulas of logical probability. Suggestions that jurors should be instructed in Bayes's theorem have not progressed far, understandably. But there is a formalism that has proved very serviceable in studying logical probability. It says:

The probability of hypothesis h on evidence e is represented by a number $P(h \mid e)$ between 0 and 1 (inclusive), which satisfies two axioms:

1. $P(\text{not-}h \mid e) = 1 - P(h \mid e)$
2. $P(h \mid h' \,\&\, e) \times P(h' \mid e) = P(h' \mid h \,\&\, e) \times P(h \mid e)$

From these axioms various theorems can be derived, such as:

- If e is a consequence of h (but not of "background evidence" b), then

  $P(h \mid e \,\&\, b) > P(h \mid b)$

  (Polya's "fundamental inductive principle" that theories are confirmed by their non-trivial consequences)

- If e is a consequence of h (but not of b) and $P(e \mid b)$ is low, then

  $P(h \mid e \,\&\, b)$ is much greater than $P(h \mid b)$

  (that is, verification of a surprising consequence renders a hypothesis much more credible)

Those are substantial reasons for accepting logical probability. They are too substantial to be dismissed with generalized complaints, for example about the difficulty of discovering the exact numerical relation between given bodies of evidence and conclusions. Objective Bayesianism does not claim that it is typically easy to discover the relation of a body of evidence to a conclusion, only that it is there to be discovered.

The theory does not in itself take a position on whether some of the evidence on which science is based is or is not certain. Simple observational facts like "I see a black bird in front of me," and easy mathematical truths like "2 + 2 = 4" have a claim to be more certain and unshakeable than complicated theories that use them as evidence. Nevertheless, it may be that a sufficiently well-confirmed theory could make us doubt apparently contradictory observations or the results of calculations. The solidity of observational and mathematical truths is something to be considered later.

Why Induction Is Logically Justified

Some people have a psychology that makes them describe as half-empty a glass that others think of as half-full. In the same way, some philosophers of science never manage to move beyond the shocking discovery that induction is fallible to ask seriously why it mostly works. True, no matter how many black ravens have been observed (without exception), it is always logically *possible* that the next one will be white. A universal generalization like "All ravens are black" is not logically implied by any finite number of observations. The same is true of any sample-to-population inference such as opinion polling: it is possible that the sample is not representative of the population. That is very old news. It does not need constant reiteration. The question is, can we rely on inductive arguments with true premises to have true conclusions *most of the time*? If I have observed many black ravens and no ravens of any other color, does that give me high confidence, or indeed any reason at all to believe, that the next one will be black?

BLACK SWANS

"All swans are white" used to be a standard example of a well-known empirical generalization as much as "all ravens are black." On January 5, 1697 a small party from Willem de Vlamingh's exploratory expedition for the Dutch East India Company landed on the coast of New Holland (near present-day Perth, Western Australia). Over the next few days they explored the mouth of a substantial river and were surprised to find swimming in it a number of birds very similar to European swans, every single one of them black. They captured a few which unfortunately died before reaching Europe, but the news necessitated a change in one of the examples which had sufficed for logicians since time immemorial. Most embarrassing.

What conclusions should we draw from this contretemps? Surely

- Induction is fallible: no amount of observational data can make an empirical generalization certain (unless we have surveyed all the cases).
- A generalization beyond the spatial or temporal range of the data (extrapolation) is less certain than one within the range of the data (interpolation).
- These sorts of events do not happen often.

Black Swan, watercolor by Richard Browne for Lieut. Thomas Skottowe's manuscript *Select Specimens from Nature* (1813). Reprinted with permission from Mitchell Library, State Library of New South Wales.

The leading argument as to why induction is justified (as a matter of logical probability) is one put forward by Donald Williams and later refined by David Stove. They explained how to reduce *inductive* inference—any inference from sample to population—to the proportional syllogism, using this argument:

- The vast majority of large samples resemble the population (in composition).
- This is a large sample.
- So, this sample resembles the population. (Equivalently, the population resembles the sample.)

This is the same kind of argument as the one considered above:

- 99.9 percent of men are mortal.
- Socrates is a man.
- So, Socrates is mortal.

The premises give good, though not conclusive, reason for believing the conclusion, as a matter of logic.

"The vast majority of large samples resemble the population" is a necessary mathematical truth, so this argument, if there is anything in it, gives an explanation of why inductive inference is generally reliable. There is no need for any "cement of the universe" such as causality or natural law to glue the unobserved to the observed, or for contingent principles about the uniformity of nature.[4]

The mathematical nature of the premise "the vast majority of large samples resemble the population" possibly needs some clarification and illustration. "Vast," "large," and "resemble" are of course imprecise words. But in whatever (reasonable) way we choose to make them precise, we arrive at a fact that can be shown to be true simply by counting.

Suppose, for example, we take a population of 100 balls, black or white in an unknown proportion, and consider samples of size 50. Let us say a sample "resembles" the population if its white/black proportion is within 4 percent of the population proportion. Thus if the actual proportion of whites is 60 percent (i.e., 60 white balls and 40 black), then a sample of size 50, which "should" have 30 whites, is said to "resemble" the population if it has 28, 29, 30, 31, or 32 whites. (A proportion of 28/50, for example, is 56 percent, which is just within 4 percent of the true 60 percent). Now—remembering that the actual population proportion of whites is unknown—what can we say about the proportion of size 50 samples that "resemble" the population?

The answer is that it is at least 68 percent. It does differ depending on the population composition—if the 100 balls are *all* white, then all the samples are all-white and so match the population exactly. But in the worst case—when half are white and half are black—68 percent of the size 50 samples resemble the population (according to our criterion: having composition within 4 percent of the population.)

Perhaps 68 percent is not quite a "vast" majority. But we did have a quite restrictive definition of "resembles," we were taking the worst case, and 50 is not a very large sample. As soon as we relax our requirements, the majority resembling the population increases.

HUME'S SKEPTICISM ABOUT INDUCTION

The contrary of every matter of fact is still possible; because it can never imply a contradiction, and is conceived by the mind with the same facility and distinctness, as if ever so conformable to reality. That the sun will not rise to-morrow is no less intelligible a proposition, and implies no more contradiction, than the affirmation, that it will rise. We should in vain, therefore, attempt to demonstrate its falsehood. Were it demonstratively false, it would imply a contradiction, and could never be distinctly conceived by the mind.

Evidence

It may, therefore, be a subject worthy of curiosity, to enquire what is the nature of that evidence, which assures us of any real existence and matter of fact, beyond the present testimony of our senses, or the records of our memory....

All reasonings may be divided into two kinds, namely demonstrative reasoning, or that concerning relations of ideas, and moral reasoning, or that concerning matter of fact and existence. That there are no demonstrative arguments in the case, seems evident; since it implies no contradiction, that the course of nature may change, and that an object, seemingly like those which we have experienced, may be attended with different or contrary effects. May I not clearly and distinctly conceive, that a body, falling from the clouds, and which, in all other respects, resembles snow, has yet the taste of salt or feeling of fire? Is there any more intelligible proposition than to affirm, that all the trees will flourish in DECEMBER and JANUARY, and decay in MAY and JUNE? Now whatever is intelligible, and can be distinctly conceived, implies no contradiction, and can never be proved false by any demonstrative argument or abstract reasoning à priori.

If we be, therefore, engaged by arguments to put trust in past experience, and make it the standard of our future judgment, these arguments must be probable only, or such as regard matter of fact and real existence, according to the division above mentioned. But that there is no argument of this kind, must appear, if our explication of that species of reasoning be admitted as solid and satisfactory. We have said, that all arguments concerning existence are founded on the relation of cause and effect; that our knowledge of that relation is derived entirely from experience; and that all our experimental conclusions proceed upon the supposition, that the future will be conformable to the past. To endeavour, therefore, the proof of this last supposition by probable arguments, or arguments regarding existence, must be evidently going in a circle, and taking that for granted, which is the very point in question.

David Hume, *Enquiry Concerning Human Understanding,* Section IV, part ii (1748).

The argument will be clarified if we understand what is wrong with David Hume's very influential argument that there cannot be any non-circular justification of inductive reasoning. After reminding us that inductions are always fallible, Hume asks what reasons there could be to connect what is observed with what is unobserved. Would that reasoning not have to be merely probable and therefore reliant on our notions of cause and effect? In that case it would rely on inductive reasoning— on an assumption that causes would continue to operate as in the past—and hence the reasoning would be circular.

The flaw in Hume's reasoning is the assumption that any probable (that is, non-deductive) reasoning must rely on knowledge of cause and effect (or some such contingent fact). That is untrue. The proportional syllogism and other standard probabilistic arguments do not rely on any assumption about cause and effect or the uniformity of nature. They are purely logical. Hume's assertion that "all arguments concerning exis-tence are founded on the relation of cause and effect; that our knowledge of that relation is derived entirely from experience" denies, without justification, that there are or even could be any probabilistic arguments that are matters of pure logic. Since it is easy to supply examples that at least appear to be exactly that, his assumption is unacceptable.[5]

THE EVILS OF INDUCTIVE SKEPTICISM

While the disablement of induction plays hob with the pedestrian undertakings of the special sciences, it will do worse with the more delicate and immense topics of metaphysics and morals. On these sciences, where sound induction is most needed and least prac-ticed, devolves the final duty of informing man where he stands and by what route he can attain salvation, in this world or another. They become pointless mummery, and the relativity of ethical judgment becomes irremediable, as soon as we dispute the ultimate validity of argument from the perceived to the unperceived.

More concretely, the implications of inductive skepticism and its eventual effects stretch down into the most intimate projects of common sense and out into the widest reaches of politics, domestic and international. If there is no rational difference between a sound scientific conclusion and the most arrant superstition, there is none between a careful investment and a profligate speculation, between a just and wise decision of a court and a flagrant miscarriage; men are hanged by a process of selection as conventional as eeny-meeny-miney-mo or the human sacrifices of the Aztecs. Indeed, there is no reason on however much theory or experience to turn the steering wheel to follow a curve in the road, or to expect gunpowder to explode, seeds to grow, or food to nourish.

In the political sphere, the haphazard echoes of inductive skepticism which reach the liberal's ear deprive him of any rational right to champion liberalism, and account already as much as anything for the flabbiness of liberal resistance to dogmatic encroachments from the left or the right. The skeptic encourages atavistic rebellion with "Who are we to say?" and puts in theory the forms and methods of democracy on a level with the grossest tyranny....

Donald Cary Williams, *The Ground of Induction* (Cambridge, MA: Harvard University Press, 1947), 18.

CHALLENGE: CONSTRUCT A NON-INDUCTIVE WORLD

If induction is justified as a matter of logic, it should be impossible to construct a toy world where it does not work. Is it possible? One's first try might be to try a chaotic world, but in that world there are no patterns so one does not make any inductive predictions (or possibly one makes some where there are small random islands of order, but in that case their predictions are often correct).

It is true that a world with a deceitful demon that changes the world specifically to confound the predictions *I* make would seem non-inductive to me, but that is cheating: a non-inductive world should be one in which *most* inductive predictions, in the abstract, are false. On reflection, it is not so easy to see how to construct one....

Central though it is, induction is only one of the components of scientific method. Before moving to some objections and then to more particular methods of knowing in the sciences—methods especially associated with one or other kind of science—we introduce briefly two general-purpose styles of argument. They are not unique to science but are pursued there so much more rigorously than anywhere else that they have a claim to constitute the heart of a special scientific method.

IS THE FAR SIDE OF THE MOON THE SAME AS THE NEAR SIDE?

Except for some small wobbles, the moon presents the same face to the earth all the time, so most of the far side is never visible. Thus for most of human history it has been a matter of pure inductive speculation as to whether the far side was like the near side. The first photographs came from the Soviet Luna 3 mission in 1959. It turned out to be much the same, except in one important respect: it has hardly any of the flat "seas" that cover over 30% of the near side.

Inference to the Best Explanation

"In the Middle Ages it was believed the earth was flat" is one of those hard-to-kill myths that infest the history of ideas. There is a whole book refuting it.[6] On the contrary, Thomas Aquinas gave "the earth is spherical" as an example of a well-known scientific truth that is not obvious, and it was a truth that had been known to all educated people since its discovery by the Greeks about 500 B. C.

It is not an easy fact to discover. It cannot be established simply by confirming instances as "all ravens are black" can, since the problem is to think of the theory in the first place, and then to deduce what observations are implied by the theory so they can be checked to confirm it. It requires an imaginative leap with geometry to put forward a new theory that explains some observed phenomena that were previously just a heap of facts. Though we do not have the account of the original discoverers, it is clear that the facts that they realized could be explained by the hypothesis of a spherical earth concerned eclipses. For example, that an eclipse of the moon occurs at full moon can be explained by supposing that a round earth comes between the sun and the moon and the eclipse is the earth's shadow passing across the moon.

Arguments of this kind, where a mass of facts is unified by a theory that explains them, are called inferences to the best explanation or abductions. Like inductions, they are non-deductive inferences, since there is no logical guarantee that the proposed explanation is the only or the best one. This form of argument is far less well-understood than deduction or induction—for example, it is unclear what it is about such theories that makes them explanatory, or why simpler theories are better explanations, other things being equal. Nevertheless, they are a crucial

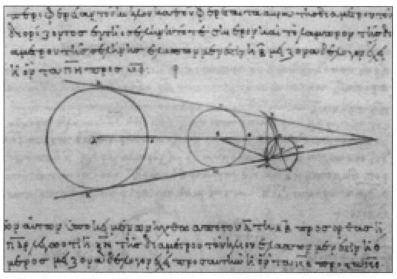

FIGURE 1.1 Aristarchus of Samos's diagram of an eclipse of the moon.

form of argument in scientific and in commonsense reasoning, and their rationality is not in question in clear cases.[7]

Being able to explain many phenomena is not in itself sufficient to make a theory well-established via an inference to the best explanation. A paranoid explanation is able to explain whatever happens, but that is a fault, since nothing counts as evidence against it. If I believe I am being followed everywhere by the CIA, and explain my failure to observe CIA agents by the fact that the CIA is very cunning, I have found an explanation for whatever I observe, but I have insulated myself from evidence against my theory. A good scientific theory must make checkable predictions as to what would happen if it were true and what (different) would happen if it were false. A relevant principle of logical probability is: "If hypothesis h is initially unlikely—on background evidence b—then a consequence of h, which is unlikely without h, increases its probability greatly."

DALTON'S ATOMIC THEORY: A BEST EXPLANATION

What is the evidence that atoms exist? The ancient Greeks and seventeenth-century philosophers proposed them as explanations of various phenomena, but could not point to any convincing cases where an explanation in terms of atoms was superior to one using the alternative and more natural hypothesis that matter is continuous. The first convincing argument was that of Dalton, who discovered instances of what is now called the law of multiple proportions—if two elements form more than one compound between them, then the ratios of the masses of the second element which combine with a fixed mass of the first element will be ratios of small whole numbers. An atomic structure of matter gives a good explanation while a continuous one has little prospect. Dalton wrote:

> *The element of oxygen may combine with a certain portion of nitrous gas, or with twice that portion, but with no intermediate quantity. In*

the former case nitric acid is the result; in the latter nitrous acid....
Nitrous oxide is composed of two particles of azote and one of oxygen.
This was one of my earliest atoms. I determined it in 1803, after long
and patient consideration and reasoning.

John Dalton, "Experimental Enquiry into the Proportion of the Several Gases or Elastic Fluids, Constituting the Atmosphere," *Memoirs of the Literary and Philosophical Society of Manchester* 6 (1805): 244–258 and citation of later lecture in A. W. Thackray, "Documents Relating to the Origins of Dalton's chemical Atomic Theory," *Memoirs & Proceedings (of the) Manchester Literary & Philosophical Society* 108, no. 2 (1966): 21–42.

Controlled Experiment

Science is rightly praised for its ability to achieve truth through observation and experiment. The two are not the same concept. Medieval science had observation but very little active experiment, leading it to be strong on topics where pure observation and analysis can achieve results, such as optics, anatomy, and the science of motion, but weak in fields where something more active is needed, such as chemistry.[8] Francis Bacon's advice that nature needed to be "tortured" to reveal her secrets, and Galileo's experiments in dynamics are two high points of the Scientific Revolution, beginning the use of active experimental methods that have continued to yield an indefinitely growing body of firmly established truths.

The essence of the experimental method, of honestly testing a putative cause against controls, is apparent in a simple paper in the *Journal of the American Medical Association*, 1998, which describes the experiment devised by Emily Rosa to test the claims of "therapeutic touch." Practitioners of the art claimed to be able to feel a tingling caused by the "energy field" surrounding a person's body when they put their hands near (not on) the body. The test had the practitioners put their hand through an opaque screen, with a coin toss selecting whether the hand of another person (invisible to them) was or was not put near theirs. Could they tell when a hand was there? They could not. Chance would have produced answers correct about 50 percent of the time, whereas the performance of practitioners was only 44 percent correct.

Miss Rosa was aged nine.[9]

It is not always as easy as that, of course, especially in the more theory-heavy sciences such as fundamental physics. In those cases, the interpretation of experiments can often depend on the results of other theories (for example, the "picture" seen in an electron microscope is a reconstruction based on a great deal of theory about how electron microscopy works). That often led to a neglect of experiment in the philosophy of science and a suspicion that there was some truth in the "Quine-Duhem thesis": that in the end experiments only confronted theory as a whole and that with enough ingenuity one could change theory to cope with the results of any experimental result whatever.

That thesis does not survive a closer look at experimental practice from a Bayesian perspective. Experiments are fallible, but it is clear why considerations such as the consistency of one set of data with other known data, repeating the experiment, carefully calibrating the instruments using theory other than the one being tested, statistical tests, and intuitions of reasonableness of the outcome compared to the predictions of theory, can act together to increase the probability of an experimental result beyond reasonable doubt.[10]

It remains true, of course, that in some sciences experiment is impossible, such as in astronomy, or unethical, such as in testing the effects of social engineering. One is then left with the difficult task of inferring causes from purely observational data. (This matter will be taken up again in chapter 11.)

We will have a lot more to say on how science knows what it does know, but this brief account is enough to give an initial account of how science can be rational—enough to defend it against its main enemies. As we will see in the next chapter, their attacks are not against any esoteric features of science, but the basic relation of evidence to conclusion itself.

CHAPTER 2

Enemies of Science
The Early Phase

Absent thee from felicity awhile. Put yourself, for the moment, in the position of one of science's enemies. You are a circa-2000 Western humanistic intellectual raised in a culture of suspicion and you are "conflicted" (in the language of psychobabble) about science. Perhaps you were not very good at it in school but have discovered a skill in spinning words. Perhaps you find science authoritarian because it is authoritative and you have a problem with authority figures. Perhaps you are indignant about Western science-based economic and military success for political reasons. Perhaps you are broke and need the money from whatever you can write fast. It doesn't matter what the causes of your animus toward science are, because you won't be telling your public about them. What you want are reasons, or strategies, to mount an attack on science. What are the resources available to you?

The first three lines of attack come courtesy of the ancient sophist Gorgias of Leontini. He achieved a lucrative *succès de scandale* in the Athens of the fifth century B. C. by defending the propositions: "Nothing exists; if anything existed, it could not be known; if anything were known, it could not be communicated." There is plenty of life still in those. So you can try any of these:

- Doubt whether there is any solid reality out there for science to know (or at least, label "naïve" the assumption that there is such a reality).
- Maintain that even if there is some sort of reality out there, we cannot know what it is because we are trapped in our own evolutionarily determined brains/cultural understandings/specific historicities/reactionary educations.
- Allege that language is incapable of communicating any truth about objective reality, as it cannot refer directly to things and their properties.

Post-Attic thought has come up with another trio of plans of attack, ones that would have seemed *outré* in the days when rationality itself was new and exciting. They are:

- Doubt the logical nature of the relation of evidence to conclusion, maintaining that the conclusions are "underdetermined" by evidence and, hence, that the gap must be filled by political decisions, wishful thinking or similar.
- Confuse ethics with logic by taking the misuse of scientific discoveries as reasons to disbelieve the discoveries themselves.
- "Put politics in charge," as Mao said: claim that the classifying of objects and the assertion of conclusions as true (why not say "absolutely true"?) is in itself authoritarian and patriarchal, hence deserving of opposition and "transgression."

You will of course try to put these together in a blender (the resulting purée has come to be called postmodernism, and has now settled in as a fixture on the intellectual scene). You will find that many have trod this path before you. That is a discovery at first exhilarating, as you observe the kudos your role models have achieved, but soon enough oppressive, as you wonder where there is space to differentiate your product and complain about something new.

Let us survey how these ideas grew over the past hundred years, on the grounds that those destined to repeat history might as well have some idea of it to begin with. You will find plagiarism more efficient than starting from scratch. Better get on with it. You have a lot of books to write.

Four Modern Irrationalists: The Attack on Evidence

The twentieth-century attack on the rationality of science came in two waves. The first, about 1935 to 1980, was spearheaded by the "four modern irrationalists" of the title of David Stove's book[1]: Popper, Kuhn, Lakatos, and Feyerabend. Taking the first of the post-Attic strategies, these four mounted a direct attack on the relation between evidence and conclusion. The second wave, which began in the English-speaking world in the disturbed times of the 1970s and whose end is frequently predicted but not yet in sight, combined a "postmodernist" strand coming from the linguistics-based humanities with a "sociology of science" strand that aimed to substitute sociology for the logic and philosophy of science.

The robustly positive tone of the title of Sir Karl Popper's massive *The Logic of Scientific Discovery*, the book that dominated philosophy of science around the 1960s, belies the fact that its thesis is that there is *no* logic of scientific discovery—or of scientific theory evaluation either. It is as if one wrote a book called *Sponges of the Aegean* to elaborate one's theory that the Aegean is a sea devoid of sponges. Or as Stove puts it, it is as if the fox in Aesop's fable, "having become convinced that neither he nor anyone else could ever succeed in tasting grapes, should nevertheless write many long books on the progress of viticulture."[2]

Yet Sir Karl genuinely believed he was a defender of science, and he found a number of followers, including some scientists, to agree with him. To appreciate later developments in antiscience, it is crucial to understand how this came about. Popper began with two convictions, the first correct and the second not. The first arose from his talking as a young man in the Vienna of the 1920s with a great variety of theorists. He was profoundly shocked by Marxists and Freudians, in particular, for pretending to "explain" with their theories whatever was observed. (Darwinians, who did the same, were spared his criticism until much later.) Surely, he said, it is not a merit of a theory to be able to give a story about whatever happens, since that means evidence cannot count against it. A genuinely scientific theory should "stick its neck out," by making predictions that could turn out to be wrong—that could be "falsified" by some possible evidence.

There is something correct in that idea, but the explication that Popper came to give of it was vitiated by his second conviction. This was that Hume had shown that, logically, inductive inference was entirely unjustified, and that there could not be anything like an inductive or probabilistic logic. Restricted to the resources of deductive logic, Popper proposed his criterion of "falsifiability": that a theory such as "all swans are white" is genuinely scientific because there are possible observations, such as of a black swan, that are strictly incompatible with it. The falsifiability criterion, he proposed, would demarcate science from non-science (Marxism, metaphysics, and so on).

Many scientists, especially those of more fertile imagination, were excited to hear of their heroic role as makers of bold and refutable conjectures. The distinguished neurophysiologist Sir John Eccles, whose theory that synaptic transmission in the brain is primarily electrical rather than chemical *had* been refuted, gratefully coauthored a book with Popper on the mind and brain. But philosophers were more skeptical. They put two lines of questioning to Popper, both related to his restriction to deductive logic. Neither problem was answered directly in Popper's later writings.

First, it was pointed out that in the real world of theory, hardly anything is strictly logically incompatible with anything else, especially with observation reports. Newton's law of gravity, normally considered confirmed by its consequences of near-elliptical planetary orbits, is in fact strictly compatible with square orbits. It just needs some complex auxiliary hypotheses about unobserved forces on the planets—unlikely hypotheses, perhaps, but "unlikely" is not allowed in Popper's theory.[3] Even if "All swans are white" is incompatible with "This is a black swan," it is not incompatible with "This appears to be a black swan," or "This is a black bird shaped like a swan," much less with "The journals of de Vlamingh's expedition report black swans on the coast of New Holland." In general, in a case of conflict between theory and observation there is often a choice of possible, if far-fetched, ways of rendering them consistent.

There are also problems for Popper with regard to statements that have more logical complexity then simple "all" statements. For example, "Every species descends from some pre-existing species" is not falsifiable in his sense because of the "some" in it: no failure to find an ancestor for a given species, or any other observation, will falsify the claim that an ancestor must exist somewhere. And, most crucially, given the statistical character of much of modern science, is the problem that "Most swans are white" or "Ninety-nine percent of swans are white" are not falsifiable either. They are logically compatible with *any* relative frequency of whiteness in observed swans.

Popper was not entirely unaware of this problem. He says that in such cases one should adopt a "methodological rule or decision to regard ... [a high] negative degree of corroboration as a falsification."[4] Plainly, to replace "p and q are consistent" with "let us regard p and q as inconsistent" is not making progress on solving the problem. Nevertheless, in proposing to replace questions of logic with sociological entities such as rules, Popper was pointing the way to the future.

A second problem for Popper is much more serious. What was his answer to the question, "If a theory has survived falsification (after rigorous testing), is it any more probable (credible, rational, believable ...) than it was before?" His answer was "no," as it had to be, given his denial of any probabilistic logic that would support the concept of an increase in probability on given evidence. In Popper's view, if a theory has survived, it is simply a survivor—that is all that can be said about it. It is not rational to believe it, or to believe it more strongly than previously, or to prefer it to any other unrefuted theory. That problem, never answered, is the one that makes Popper the true godfather of the irrationalist camp.

That still leaves some explanation needed for what was right about Popper's original insight: that making risky predictions is a good thing in a scientific theory. Objective Bayesianism has an answer to that, namely, that if a theory makes a risky prediction which then turns out to be true, the theory becomes more probable than it was before (and the more risky the prediction, the more probable the theory becomes). That is a theorem of (logical) probability:

If $P(h \mid b)$ is low, and h & b implies e, while $P(e \mid \text{not-}h$ & $b)$ is low, then $P(h \mid e$ & $b)$ is much higher than $P(h \mid b)$

("If hypothesis h is initially unlikely—on background evidence b—then a consequence of h, which is unlikely without h, increases its probability greatly.")

It is certainly ironic that Popper, avowed opponent of probability, piggybacked to fame on a single theorem of logical probability. Surely all philosophers who have discussed Popper with scientists have found that the scientists believe Popper's contribution was to show that one cannot be *certain* of universal generalizations on the basis of any finite amount of observation, and to urge that tests should be severe. If it is further pointed out to the scientists that everyone agrees with those theses, and that Popper's contribution was in fact to say that the theories are no more probable even after severe testing, that is treated as a minor technicality of little interest. The tendency of scientists to baptize principles of logical probability with Popper's name is well illustrated by a remark of Eccles, one of the most vociferously Popperian of scientists:

> *Often I have only a very vague horizon-of-expectations (to use Popper's felicitous phrase) at the beginning of a particular investigation, but of course sufficient to guide me in the design of experiments; and then I am able to maintain flexibility in concepts which are developed in the light of the observations, but always of a more general character so that the horizon-of-expectations is greatly advanced and developed and itself gives rise to much more rigorous and searching experimental testing....[5]*

And the immunologist Sir Peter Medawar, author of the much-quoted remark, "I think Popper is incomparably the greatest philosopher of science that has ever been,"[6] undermined his adulation with his account of what he thought Popper said:

> *Popper takes a scientifically realistic view: it is the daring, risky hypothesis, the hypothesis that might so easily not be true, that gives us special confidence if it stands up to critical examination.[7]*

The dispute about Popper and the logic of science remained largely confined to the world of philosophy and the musings of retired scientists. Questions about the rationality or otherwise of scientists were brought to a wider intellectual audience—

indeed, the widest possible intellectual audience—through the efforts of Thomas Kuhn.

The Kuhnian Paradigm of Antiscience

For an insight into trends and fads in the humanities world, it is hard to go past the *Arts and Humanities Citation Index*. It lists all citations in the major humanities journals—that is, an army of trained slaves keys in every footnote of every article, and the computer rearranges them according to the work cited. The compilers of the *Index* examined the records for the years 1976–1983, and issued a report on the most cited works written in the twentieth century. The most cited *author* was Lenin, which speaks volumes on the state of the humanities in the West toward the end of the Cold War. But the most cited single works were, in reverse order: in third place, Northrop Frye's *Anatomy of Criticism*, second, James Joyce's *Ulysses*, and, well in the lead, Thomas Kuhn's 1962 book, *The Structure of Scientific Revolutions*.

Interest in Kuhn's book has not waned. The *Index* is now online, and 554 citations are recorded in the Indexes for all disciplines (Arts and Humanities, Social Science and Science) for 2007. To call the tone of most of these citations reverential would be an understatement. It is reported that *Structure* is Al Gore's favorite book. Safire's *New Political Dictionary* has an article on "paradigm shift"—a phrase popularised by Kuhn—that reports both George Bush senior and Bill Clinton being much impressed with its usefulness. In very recent years, commentators as diverse as Clifford Geertz and George Soros have rushed into print on the book's importance.

The basic content of Kuhn's book can be inferred simply by asking: What would the humanities crowd *want* said about science? Once the question is asked, the answer is obvious. Kuhn's thesis is that scientific theories are no better than ones in the humanities. The thesis that science is all theoretical talk and negotiation that never really establishes anything is one that caused trouble long ago for Galileo, who wrote:

> *If what we are discussing were a point of law or of the humanities, in which neither true nor false exists, one might trust in subtlety of mind and readiness of tongue and in the greater experience of the writers, and expect him who excelled in those things to make his reasoning more plausible, and one might judge it to be the best. But in natural sciences whose conclusions are true and necessary and have nothing to do with human will, one must take care not to place oneself in the defense of error; for here a thousand Demostheneses and a thousand Aristotles would be left in the lurch by every mediocre wit who happened to hit upon the truth for himself.*[8]

Kuhn's "achievement" was to put the view of Galileo's scholastic opponents back on the agenda. Up to his time, philosophy of science had concentrated on such questions as how evidence confirms theories and what the difference is between science and pseudoscience, that is, questions about the logic of science. Kuhn declared logic outmoded and replaced it with history.

A caricature of his opinions is this: A science, say astronomy, is dominated for a long period by a "paradigm," such as Ptolemy's theory that the sun and planets revolve around a stationary earth. Most work is on "normal science," the solving of standard problems in terms of the reigning paradigm. But anomalies—results that the paradigm cannot explain—accumulate and eventually make the paradigm unsustainable. The science enters a revolutionary phase, as a new paradigm such as Copernicus's heliocentrism comes to seem more plausible. Defenders of the old order, who cannot accommodate the change, and usually cannot even understand the concepts in which it is expressed, gradually die out and the new paradigm is left in control of the field. Then the process repeats. According to the summary in Fukuyama's *The End of History*,

"The cumulative and progressive nature of modern science has been challenged by Thomas Kuhn, who has pointed to the discontinuous and revolutionary nature of change in the sciences. In his most radical assertions, he has denied the possibility of 'scientific' knowledge of nature at all, since *all* 'paradigms' by which scientists understand nature ultimately fail."

As with many caricatures, one finds that the original consists of the caricature with the addition of a number of qualifications that render it inconsistent, the number of those inconsistencies multiplying with the author's subsequent denials that he meant to say anything so crude. One observes also that the caricature has a historical career considerably more vigorous than the original, whose qualifications would have lessened its appeal.

Besides its simplicity, Kuhn's caricature makes the story of science into one of the simple, emotive plotlines that literary folk find so engaging. It is the story of *Morte d'Arthur*, of the peaceable order and its aging king, their virtue undermined by internal corruption, falling to the challenge of the vigorous and bloodthirsty young upstart. The plot had made Frazer's *Golden Bough,* with its stories of tribal chiefs displacing one another with extreme prejudice, a literary hit decades before, and had even persuaded the humanities world to take an interest in the doings of Red Deer, among whom the transfer of harems between dominant males is conducted on similar principles.

Kuhn's success is also an instance of the enduring appeal of *theomachy,* a mode of explanation that worked so brilliantly for Marx and Freud, and, long before, for Homer. What was previously thought to be a continuous and uninteresting succession of random events is discovered to be a conflict of a finite number of hidden gods (or classes, complexes, paradigms, and so on, as the case may be), which manipulate the flux of appearances to their own advantage, but whose machinations may be exposed by the elect to whom the interpretative key has been revealed.

Further reasons for Kuhn's success are not hard to find. He gave permission to anyone who wished to comment on science to ignore completely the large number of sciences that undeniably are progressive accumulations of established results—sciences such as ophthalmology, oceanography, operations research, and ornithology, to keep to just one letter of the alphabet. That certainly saved a lot of effort. Kuhn's theory had a special appeal to social scientists as well. Political scientists, sociologists, and anthropologists recognized Kuhn's picture of disciplines relegating the accumulation of evidence to the background while bringing fights about theory to the fore; they were delighted to hear that what they had until then considered an embarrassment was the way things were done in the most respectable sciences. Kuhn even offered something to massage the egos of natural scientists themselves. It might seem at first glance that his claim that most scientists are drones was an insult, but there was a good reason why it was met with the same equanimity one notices in fundamentalist religious circles at the news that only 144,000 are saved. The damned may be a majority, but of course they are *other* people; every scientist had the opportunity to cast himself as a revolutionary hero of a new paradigm, shamefully ill-used by the establishment.

Kuhn's rhetoric incorporated a few further successful ploys, in that "paradigm" was undoubtedly a cute technical term, as technical terms go, and the phrase "normal science" had just the right hint of superciliousness toward the worker bees who credulously do the hard work of science. Kuhn's work was the perfect product of the sixties, and since he managed to publish it in 1962, his success was inevitable—indeed, as the philosophers say, overdetermined.

At a more logical level, Kuhn's success depended on certain ambiguities. Even in the caricature above, it is clear how some of these were essential to Kuhn's plan. What does "unsustainable" mean, as said of a scientific theory facing anomalies? In particular, is it a matter of logic or of psychology? If it means that there are a number of observed results that would be unlikely if the theory were true, then one is back in the realm of logic, of the bad old philosophy of science that studied the relation between evidence and hypothesis. Naturally, Kuhn was not keen to emphasize that direction. But if "unsustainable" is a purely psychological matter, a kind of collective disgust by a *salon des refusés* of younger scientists who simply think their elders are too smug, then it is impossible to see why it should have any standing as science. If the old theory is not broke—if its predictions are true, for example, and its explanations coherent—why fix it? Whatever there is to be said for a pure appetite for novelties in the art world, there is no scope for it in science. There, the difficulty of attaining the truth means no one is inclined toward pointless exercises in throwing away pearls attained at great expense.

After Kuhn

No later opponent of science matched Kuhn in popularity. Imre Lakatos's work dealt mainly with mathematics and will be considered later, though his more general work on the "methodology of scientific research programmes" was notable for its persistent use of quotation marks to undermine claims to knowledge (as in "One typical sign of the degeneration of a programme ... is the proliferation of contradictory 'facts.' Using a false theory as an interpretative theory, one may get—without committing any 'experimental mistake'—contradictory factual propositions, inconsistent experimental results."[9])

Feyerabend, the last of the "four irrationalists," was an *enfant terrible* whose "Anything Goes" comic term, according to which voodoo is as rational as science, was an embarrassment to his own side. Defenders of science enjoyed exhibiting him as a *reductio ad absurdum* of the irrationalist trend, while Popper and Kuhn faced the difficult task of explaining why their views did not imply his. His opinions are perhaps not shared widely enough to make analysis of them worthwhile.

Something similar is true of the British continuation of the irrationalist trend in the "strong program" in the sociology of science that is associated with David Bloor, Barry Barnes, and Steve Woolgar. Taking Kuhn's sociological approach to an extreme, these three proposed to completely replace the philosophy of science and its disputes about logic and method with a sociological approach, which would examine communities of scientists and their decision-making without any commitment to the truth of their theories, much as an anthropologist would consider the beliefs of a primitive tribe.[10] The project of trying to understand why scientists came to believe theories without alluding to the evidence they had for them proved easier said than done. Four philosophers lived with a group at a leading medical research institute in Melbourne that worked on malaria vaccine and observed their behavior from an anthropological point of view. The resulting book did not work out quite according to plan. Some of the authors thought the scientists were engaged in exactly the construction of fictions expected, but others came away with the impression that the researchers were actually discovering things about immunology and malaria. The book had to go to print with a disappointingly noncommittal conclusion.[11]

The arguments against Kuhn's half-hearted embrace of sociology apply *a fortiori* against all subsequent full-blooded sociological approaches. The front had moved elsewhere. The French had arrived.

HELP TO YOUNG AUTHORS

Neutralizing success words, after the manner of the best authorities.

How to rewrite the sentence: Cook discovered Cook Strait.

Lakatos:

Cook "discovered" Cook Strait.

Popper:

Among an infinity of equally impossible alternatives, one hypothesis which has been especially fruitful in suggesting problems for further research and critical discussion is the conjecture (first "confirmed" by the work of Cook) that a strait separates northern from southern New Zealand.

Kuhn:

It would of course be a gross anachronism to call the flat-earth paradigm in geography mistaken. It is simply incommensurable with later paradigms: as is evident from the fact that, for example, problems of antipodean geography could not even be posed under it. Under the Magellanic paradigm, however, one of the problems posed, and solved in the negative, was that of whether New Zealand is a single land mass. That this problem was solved by Cook is, however, a vulgar error of whig historians, utterly discredited by recent historiography. Discovery of the Strait would have been impossible, or at least would not have been science, but for the presence of the Royal Society on board, in the person of Sir Joseph Banks. Much more research by my graduate students into the current sociology of the geographical profession will be needed, however, before it will be known whether, under present paradigms, the problem of the existence of Cook Strait remains solved, or has become unsolved again, or an un-problem.

Feyerabend:

Long before the constipated and boneheaded Cook, whose knowledge of the optics of his telescopes was minimal, rationally imposed, by means of tricks, jokes, and non-sequiturs, the myth of Cook Strait on the "educated" world, Maori scientists not only "knew" of the existence of the Strait but often crossed it by turning themselves into birds. Now, however, not only this ability but the very knowledge of the "existence" of the Strait has been lost forever. This is owing to the malignant influence exercised on education by authoritarian scientists and philosophers, especially the LSE critical rationalists, who have not accepted my criticisms and should be sacked. "No doubt this financial criticism of ideas will be more effective than [...] intellectual criticism and it should be used."

Boston Studies in the Philosophy of Science 43 (1978): 144 from David Stove *Anything Goes: Origins of the Cult of Scientific Irrationalism* (Popper and After, Pergamon Press, 1982; reprint, Macleay Press, 1998), ch. 1.

CHAPTER 3

Enemies of Science
The Postmodernist Phase

Postmodernism is not so much a theory as an attitude. It is an attitude of suspicion—suspicion about claims of truth. So if postmodernists are asked, "Aren't the claims of science just true, and some things objectively right and wrong?" their reaction is not so much "No, because ...," but "They're always doubtful, or relative to our paradigms, or just true for dominant groups in our society, and anyway, in whose interest is it to think science is true?"

Postmodernism is an attitude not only of suspicion, but of *unteachable* suspicion. If one tries to give good arguments for some truth claim, the postmodernist will be ready to "deconstruct" the concept of good argument, as itself a historically conditioned paradigm of patriarchal Enlightenment rationality.

Not only that, the postmodernist *congratulates* her/himself morally on being unteachably suspicious. Being "transgressive" of established standards is taken to be good in itself and to position the transgressor as a fighter against "oppression," prior to giving any reasons why established standards are wrong. In asking how to respond to postmodernism, it is especially important to understand that its motivation does not lie in argument but in the more primitive moral responses, resentment and indignation.

Postmodernist "Arguments"

Actual arguments against rationality in science or elsewhere, therefore, are not essential to the postmodernist project. Indeed, they are considered as being collusive in rationalist oppression by granting too much to the other side (as if winning or losing the argument would be to the point). It is therefore not easy to isolate postmodernist arguments so as to analyze them, nor is it clear if the effort to do so is worthwhile, given their subsidiary role in the performance. Nevertheless there is an argumentative core to the postmodernist rejection of rationality, and it is possible to say what is wrong with it.

In summary, it is Gorgias of Leontini over again. But not so concise.

There are obvious difficulties with presenting the arguments in the original works of Derrida, or Lacan, or Baudrillard. They do not write in any natural language, they do not put the premises before the conclusion, the conclusion is distributed over the text rather than appearing in any one sentence, positions are assumed to have been established outside the texts one is actually reading, in previous texts, or perhaps future ones, and so on. But a broadly accurate summary of their most basic argument is given by an author who writes in English (or almost):

> *Based on the Saussurean principle of the sign, which is that the relationship between the signifier ... and the sign ... is arbitrary, the structure of language for Lacan is such that "language" is already cut off from "reality." What is taken as the meaning ... of any word, for example, is always going to be the result of that word's difference from all other words within a particular language. Meaning, then is a result of difference, and difference is a result of language as a system.... Consequently the Saussurean-*

*based theory of language ... is radical because it erases "reality"
from the system: reality is never present "in" or "to" the system of
language.... The gap between word and thing ... is a necessary
one inasmuch as language can never be identical with what it
names, for example, and vice versa.... From this it follows that
presence (truth, reality, self-identity) is an effect of a system that is
constituted by absence and separation. The very lack within language and the very gap between word and thing is what makes
reality possible, making it seem present.[1]*

The postmodernist attack is, then, based on a thesis about language. That is, it is a version of Gorgias's third assertion: "If anything were known, it could not be communicated." As exposed
in Raymond Tallis's brilliantly titled *Not Saussure*,[2] it depends
on an implausible thesis of the French theorist of language Fer-
dinand de Saussure. His thesis was that language cannot genuinely refer to reality but only to other language—that "black"
gets its meaning solely as the opposite of "white" (and its relation to other words in the language) rather than by referring to
black things because of having a learned connection with them.
Saussure's "structuralist" theory of language encourages concentration on the internal systematics of language rather than on
its basic purpose: to mean something outside itself. As applied
in "poststructuralist" and postmodern "discourse," it acts as a
universal solvent of claims to truth, scientific or other. As with
the move from logic to sociology, the purpose of this ploy is to
undercut any talk about people believing something because
there is good reason to, thus opening the way to endless "decon-
structive" speculation on the political and psychopathological
causes of their belief.

Adopting a policy of deliberate unteachability on the relation between reasons and belief, or between language and the
world, risks not only the destruction of the critical faculties
of one's students but, in due course, the atrophy of one's own.
Those critics who braved the dense thickets of prose of the

postmodernist writers came to suspect they could not tell the difference between what they were writing and deliberate nonsense. It occurred to one scientist that this was a hypothesis susceptible of scientific testing....

The Sokal Hoax

Alan Sokal, a physicist at New York University, wrote a spoof article, "Transgressing the Boundaries: Toward a Transformative Hermeneutics of Quantum Gravity," containing gobbets of postmodernist nonsense that parodied what "cultural theorists" had been writing about science.

SOKAL'S HOAX ARTICLE "TRANSGRESSING THE BOUNDARIES"

There are many natural scientists, and especially physicists, who continue to reject the notion that the disciplines concerned with social and cultural criticism can have anything to contribute, except perhaps peripherally, to their research. Still less are they receptive to the idea that the very foundations of their worldview must be revised or rebuilt in the light of such criticism. Rather, they cling to the dogma imposed by the long post-Enlightenment hegemony over the Western intellectual outlook, which can be summarized briefly as follows: that there exists an external world, whose properties are independent of any individual human being and indeed of humanity as a whole; that these properties are encoded in "eternal" physical laws; and that human beings can obtain reliable, albeit imperfect and tentative, knowledge of these laws by hewing to the "objective" procedures and epistemological strictures prescribed by the (so-called) scientific method....

In this way the infinite-dimensional invariance group erodes the distinction between observer and observed; the π of Euclid and the G of Newton, formerly thought to be constant and universal, are now perceived in their ineluctable historicity; and the putative observer becomes fatally de-centered, disconnected from any epistemic link to a space-time point that can no longer be defined by geometry alone....

More recently, Lacan's *topologie du sujet* has been applied fruitfully to cinema criticism and to the psychoanalysis of AIDS. In mathematical terms, Lacan is here pointing out that the first homology group of the sphere is trivial, while those of the other surfaces are profound; and this homology is linked with the connectedness or disconnectedness of the surface after one or more cuts. Furthermore, as Lacan suspected, there is an intimate connection between the external structure of the physical world and its inner psychological representation *qua* knot theory ... catastrophe theory, with its dialectical emphasis on smoothness/discontinuity and metamorphosis/unfolding, will indubitably play a major role in the future mathematics; but much theoretical work remains to be done before this approach can become a concrete tool of progressive political praxis....

Alan D. Sokal, opening and selections of "Transgressing the Boundaries: Toward a Transformative Hermeneutics of Quantum Gravity," *Social Text* 46/47, (spring/summer 1996): 217–252.

Anyone who knew even high school science, or had a basic feel for the relation between evidence and hypothesis, should have been able to see that the Sokal article was a spoof.

He submitted it to *Social Text*, a leading American journal of cultural studies that had advertised a special issue on the "Science Wars." The journal fell for it and printed it (no. 46/7, Spring/Summer, 1996), whereupon Sokal owned up and a good laugh was had by all. Except, of course, the editor of *Social Text* and all of like mind.

Debate raged. Sokal's website on the hoax includes almost two hundred items including one that made it to the front page of the *New York Times*.[3]

The mode of thought that Sokal was parodying, is, as even a few extracts from his article show, a wide-ranging one. It churns together pieces of political jargon, psychoanalysis, physics, mathematics, and any intellectual *objets trouvés* to hand. But its central doctrine is much the same as that of the "strong programme": It is the social construction of science, and of knowledge generally; the objects of knowledge do not have objective reality "out there," but are merely social constructs.

Stove and the Worst Argument in the World

At the back of all the efforts to undermine scientific knowledge from sociological and linguistic standpoints are inchoate versions of an argument named by David Stove as the winner in his 1985 "Competition to Find the Worst Argument in the World."

In Stove's marking scheme, half the marks went to the degree of badness of the argument, half to the degree of its endorsement by philosophers. Thus an argument was sought that was both very bad and very prevalent. He awarded the prize to himself, for the following argument.[4]

We can know things only:

- as they are related to us
- under our forms of perception and understanding
- insofar as they fall under our conceptual schemes, etc.

So, we cannot know things as they are in themselves.

Perhaps that argument does not look familiar at first glance. It is extraordinarily common, and it has underpinned many irrationalist programs in the history of thought, from classical idealism to recent relativisms in the philosophy of language, the philosophy of science, ethics, and elsewhere.

To start close to home: Two short passages from Stove's later book, *The Plato Cult*, deal with people everyone has actually met. Speaking of the typical products of a modern high school, he writes:

Their intellectual temper is (as everyone remarks) the reverse of dogmatic, in fact pleasingly modest. They are quick to acknowledge that their own opinion, on any matter whatsoever, is only their opinion; and they will candidly tell you, too, the reason why it is only their opinion. This reason is, that it is their opinion.[5]

And who can fail to recognize Stove's picture of another group of players in the intellectual world?

The cultural-relativist, for example, inveighs bitterly against our science-based, white-male cultural perspective. She says that it is not only injurious but cognitively limiting. Injurious it may be; or again it may not. But why does she believe that it is cognitively limiting? Why, for no other reason in the world, except this one: that it is ours. Everyone really understands, too, that this is the only reason. But since this reason is also generally accepted as a sufficient one, no other is felt to be needed.[6]

These arguments—or, less euphemistically, dogmas—are versions of Stove's "Worst Argument" because all there is to them as arguments is: Our conceptual schemes are *our* conceptual schemes, so, we cannot get out of them (to know things as they are in themselves). Or, "We have eyes, therefore we cannot see."[7]

The basic postmodernist argument quoted above using Saussure is a linguistic version of the Worst Argument. It says, "We cannot speak about things except through the forms of language, therefore we cannot speak about things as they are in themselves." Kuhn hints at such an argument, but no more, in comments such as "There is, I think, no theory-independent way to reconstruct phrases like 'really there'; the notion of a match between the ontology of a theory and its 'real' counterpart in nature now strikes me as illusive in principle. Besides, as a historian, I am impressed with the implausibility of the view."[8]

His followers have made up the slack, especially those in the "Strong Programme in the Sociology of Knowledge" or social constructivists such as Bloor. They propose to replace all considerations of logic, of what scientific theories are reasonable, with considerations of sociology, that is, of what interests a theory serves. The real reason for their views is the conviction that since science is done by people, its explanation should be in the realm of causes acting on people, not the realm of abstract reasons. People, they think, can be acted on by their interests, or patronage, or the social milieu, but abstract facts like $2 + 2 = 4$ do not act. So explanations of how people, including scientists, think ought to be sociological. This argument appears in various forms, mostly not very explicit ones. Thus, Bloor argues, observation "underdetermines" theory—that is, that several theories are logically compatible with any given body of observations—and concludes immediately that social factors are what must determine which theory is chosen.[9] He says that the "existence of nature" does not account for (scientific) theories and that simple "attention to nature" will not adjudicate the merits of our theories.[10] He reserves particular anger for the opinion that belief in reasonable theories is at least in part explained by their being reasonable, while mistakes require causal explanations. This, Bloor says sarcastically, is an attempt to render science "safe from the indignity of empirical explanation." Barnes and Bloor write:

Our equivalence postulate is that all beliefs are on a par with one another with respect to the causes of their credibility. It is not that all beliefs are equally true or equally false, but that regardless of truth and falsity the fact of their credibility is to be seen as equally problematic. The position we shall defend is that the incidence of all beliefs without exception calls for empirical explanation and must be accounted for by finding the specific, local causes of this credibility.[11]

It must be emphasized that Bloor does not admit any possibility of cooperation between causes and reasons: explanation in terms of causes is quite different from that in terms of reasons, he says; if one is right, the other is wrong.[12]

This argument, the central plank of the social constructivist position, is a version of Stove's "Worst Argument" because it says: "We can know things only via causal (social) processes acting on the brains of real scientists, therefore, the content of our theories is explained without remainder by the social factors causing them; that is, we cannot know things as they are in themselves." This is why no amount of raging about relativism, skepticism, and truth is found to make any impact on constructivists. They have a last line of defense in the argument: "Those entities in Platonic worlds, like truths and theories, cannot cause belief in themselves. Scientists are *people*, after all, and as such are responsive only to social or similar causes."

Like all such arguments, Bloor's says, in effect, that the mere fact that a theory is accepted is a reason for not accepting it.

After Sokal: Bourbon Antiscience

It was said of the restored Bourbons, possibly by Talleyrand, that they had learned nothing and forgotten nothing. The postmodernists' state of being "in denial" over the Sokal hoax began when the editor of *Social Text*, Andrew Ross, published the other articles in the issue to which Sokal had contributed, in a book titled *Science Wars*. Sokal's name does not appear in the index of the book, but a paragraph in Ross's introduction regrets that some may have taken Sokal's "faux version of science studies" seriously but that "it also brought a much wider audience to the topics addressed by science studies." Ross was particularly hurt that some "declared progressives" had joined in the denunciation of those of his own camp for "straying from the paths of righteousness."[13] (This was a reference to Sokal's moderately leftist views on such issues as Nicaragua.)

Sokal very generously, with a collaborator Jean Bricmont, wrote a book titled *Intellectual Impostures: Postmodern Philosophers' Abuse of Science*[14] that painstakingly went through the misapprehensions about science in the texts of the leading postmodernists, and provided a brief tutorial on why science should be believed. The postmodernists avoided citing it.

Instead of considering Sokal's criticism, the postmodernists looked for more radical ways to undermine the relation of thought to reality. One popular French guru was Gilles Deleuze, who went back for another bite of the Gorgian cherry with the thesis that if anything did exist, it didn't exist as a unity. He proposed "a philosophical ontology of Being as pure difference or becoming. Being, for Deleuze and Guattari, is that which differs from itself, in nature, always already, in itself, qualitatively different."[15] "Deleuze and those of his generation sought to conceptualise both difference and becoming, but a difference and becoming that would not be the becoming *of* some being."[16] That invites one to congratulate oneself on "resisting" totalizing discourses such as science that naively speak as if there are continuing objects with persistent properties. As soon as anyone asserts that object A has property B, the Deleuzian is ready to abuse the notions of both solid object A and continuing property B. Instead of things and properties there are "virtualities," "potentials," and "becomings." Naturally, that means that any discussion of scientific truth will be negative, and any scientific terminology used in it will probably be … well, as in this example, from Deleuze's translator and English-language interpreter Brian Massumi:

> Just as "higher" functions are fed back—all the way to the subatomic (that is position and momentum)—quantum indeterminacy is fed forward. It rises through the fractal bifurcations leading to and between each of the superposed levels of reality. On each

level, it appears in a unique mode adequate to that level. On the level of physical macrosystems analyzed by Simondon, its mode is potential energy and the margin of "play" it introduces into deterministic systems (epitomized by the "three body problem" so dear to chaos theory). On the biological level, it is the margin of undecidability accompanying every perception, which is one with a perception's transmissibility from one sense to another. On the human level, it is that same undecidability fed forward into thought, as evidenced in the deconstructability of every structure of ideas (as expressed, for example in Gödel's incompleteness theorem and in Derrida's différance). Each individual and collective human level has its own peculiar "quantum" mode; various forms of undecidability in logical and signifying systems are joined by emotion on the psychological level, resistance on the political level, the spectre of crisis haunting capitalist economies.... The use of the concept of the quantum outside quantum mechanics, even as applied to human psychology, is not a metaphor....[17]

The Citation Indexes report enthusiastic take-up of Massumi's ideas. Recent articles citing the book that contains this quote include such titles as "Aesthetics of Intermediality" (*Art History,* June 2007), "Deleuze and Space" (*Annals of the Association of American Geographers,* Dec. 2007), "Caught in the Terrains: An Inter-referential Inquiry of Trans-border Stardom and Fandom" (*Inter-Asia Cultural Studies,* 2007), and "Broccoli and Desire" (*Antipode,* Nov. 2007).

This tradition has one message to the public: "We can't learn. We won't learn (but please, keep our academic salaries coming)."

CHAPTER 4

The Furniture

B efore we can get to science proper, we must have a grasp of the concepts science uses, concepts that by and large it shares with common sense. As has become evident from the ham-fisted efforts of Artificial Intelligence to imitate it, common sense is a very complex, subtle, and powerful instrument for knowing the world.

A World of States of Affairs
According to traditional elementary-school readers,

> The cat sat on the mat.

That is a statement on which the findings of science can cast light, explaining, for example, why it is characteristic of cats to prefer mats to floors. But understanding the statement itself requires a grasp of many categories that are both prior to science and essential to and assumed by it.

"The," in "the cat," indicates that we are dealing with a single unified *object*, cut out from the background. In the continuum of matter that is the universe and the flux it undergoes, who cut out this warm furry item, drew its boundaries, and pointed it out as an individual thing deserving a common noun?[1] Although science on the smallest and largest scales may be somewhat vague on individual objects, identifying those objects in the scale where cats live ("medium-sized dry goods," as the philosophers say) is not a mere reflection of human interests—it is a factual matter that when the cat moves from the mat to the food bowl, its tail goes with it, a result of the naturalness of the unity of the warm furred parts.

That "cat" is a *common* noun is a sign of another natural unity, that between a cat and its conspecifics. That similarity too is independent of human interests, being recognized by cats themselves when negotiations over territory take place. Science seeks naturalness in classification, a way of dividing objects into classes that will put together those which by nature are really similar. (It is somewhat different with "the mat," where both the decision to regard a piece of material as a single object and its classification as a kind of artifact distinct from, for example, a cushion are to some degree reflective of human interests. As Aristotle says, the substance of a bed is not "bed" but wood, and for the same reason there is not exactly a science of mats, though there is one of textile technology.)

"Sat" is a verb of *action*, a word that one would not use literally of a stone positioned on a mat. Actions form a category reserved to kinds of things that have at least a minimal capacity for decision, and hence are possessed of some degree of mental life. How far down the chain of animal species mental life exists is a difficult question on which science has cast some light, but the basic distinction between mental and non-mental is one with which we have direct pre-scientific acquaintance, being mental ourselves. "Sat" is also causal, in that the action taken is the *cause* of the subsequent relation of sitting.

And "sat" is tensed, indicating the basic category of time in which *events* or *processes* such as sitting occur (or possibly, which is created by events or processes). "On" indicates a spatial relation. Both spatiality and relationality are essential to science. Physical things are all in a space—as far as is known, all in the same space—which at least in our region and on our scale is three-dimensional, nearly flat, and divisible at least as far as the limits of measurement. (More on space and time later.)

Finally, the combination into which "the cat," "the mat," and "sat on" enter is the uniquely possible one. We live in a "world of states of affairs,"[2] such as "The cat sat on the mat," or "The minimum temperature at the North Pole on January 3, 1907 was –33°C," or "Life first existed on earth about 4 billion years ago," or "Hydrogen is lighter than oxygen," in which particulars have properties or stand in relation to other particulars. Science aspires to lay down a "view from nowhere"[3] that describes and understands these states of affairs and their necessary interconnections.

The Categories

At the highest level, what kinds are there? Not just what kinds of *things* (animal, vegetable, mineral), but what kinds of everything real, properties as well as things. The parts of speech in natural language are an initial guide—if nouns name things (individuals and kinds, concrete and abstract), what about the actions described by verbs, the qualities described by adjectives and the relations described by prepositions? All of them are aspects of reality, surely?

Aristotle had a suggestion of ten categories, the basis of all discussion since. By and large, the language of present-day science fits in just as well with those categories as does natural language.[4]

ARISTOTLE'S CATEGORIES

1. Substance (individual things or stuff, or kinds of them)
2. Quantity

3. Quality

4. Relation

5. Place

6. Time

7. Position (i.e. the relative position of the parts of the object)

8. State (e.g. clothed or sleeping)

9. Action

10. Affection (being affected)

Aristotle, *Categories*, Section 1.4.

CATEGORICAL IMPOSSIBILITIES GRAMMATICALLY POSSIBLE

1. Colorless green ideas sleep furiously.

2. Furiously sleep ideas green colorless.

It is fair to assume that neither sentence (1) nor (2) (nor indeed any part of these sentences) had ever occurred in an English discourse. Hence, in any statistical model for grammaticalness, these sentences will be ruled out on identical grounds as equally "remote" from English. Yet (1), though nonsensical, is grammatical, while (2) is not.

Noam Chomsky, *Syntactic Structures* (S'-Gravenhage: Mouton, 1957), 15.

One would expect that science might cast some light on the list of categories, or at least not be constrained to accept uncritically the "folk categories" of natural language. Although the categories are so general that it is very hard for empirical evidence to bear on them, there have been some rare cases where experimental evidence indicated that something was in a different category from the one it was thought to be in. In the eighteenth century, the leading theory of heat held that it was a sort of fluid called "caloric," that is, in the category of substance.[5] Rumford and Joule's cannon-boring experiments showed that it was not, but must be something more like work or action.

RUMFORD'S CANNON-BORING EXPERIMENT:
EXPERIMENTAL EVIDENCE AS TO WHAT CATEGORY
HEAT IS IN

What is heat? Is there any such thing as an *igneous fluid*? Is there anything that can with propriety be called *caloric*?

We have seen that a very considerable quantity of heat may be excited in the friction of two metallic surfaces and given off in a constant stream or flux, *in all directions*, without iteration or intermission, and without any signs of diminution or exhaustion....

And, in reasoning on this subject, we must not forget to consider that most remarkable circumstance, that the source of the heat generated by friction, in these experiments, appeared evidently to be *inexhaustible*.

It is hardly necessary to add that anything which any *insulated* body, or system of bodies, can continue to furnish *without limitation* cannot possibly be a *material substance*: and it appears to me to be extremely difficult, if not quite impossible, to form any distinct idea of anything, capable of being excited and communicated, in the manner the heat was excited and communication in these, except it be MOTION.

Benjamin Thomson, Count Rumford, "An Experimental Enquiry Concerning the Source of the Heat which Is Excited by Friction," *Philosophical Transactions of the Royal Society* 88 (1798): 80–102.

Categorical problems have also been significant for the long-running disputes over whether light was made of particles or waves. Particles are in the category of substance, whereas waves are motions, and thus in the category of action (in Aristotle's scheme, at least). Yet, experiments have favored one or the other theory, or, as it turned out, some merger of the two. A more recent example of scientific uncertainty as to category is the search for "florigen." Plants "know" to flower, in normal circumstances, when their leaves (not the stem tip that flowers) are exposed to increasing day length. Experiments in the 1930s showed that a graft from an "induced" plant could also induce flowering in a plant that had not been exposed to extra light. It was concluded that a substance dubbed "florigen" was created in the leaves by an increase in light and then transmitted to the flowering parts, and the search began to identify the chemical composition of this substance. Sixty years of effort failed to find any such substance, and consensus was reached that there was no such thing as florigen and that the phenomenon of flowering would need to be explained in terms of some other category, perhaps complex ratios of substances. Some recent research has revived the substance theory, but the problem has not reached a resolution.[6]

In recent times, actual research on categories has mostly proceeded in the subfield of Artificial Intelligence that deals with understanding natural language. The discipline of formal ontology looks for an appropriate "upper level ontology" that will organize everything the language talks about into its basic categories, which will help determine, for example, whether assertions are meaningless or not. Research is active and agreement has not yet been reached.[7]

Classification

Even if the categories are understood, there is much still to discover about classification in the ordinary sense, which determines the kinds of things (in the category of substance).

It is all very well to use "All As are Bs" as a schema of induction and discuss the rationality of science in terms of the evidence for such propositions, as we did in chapter 1, but how is one to obtain the concepts A and B? Might it have been possible, for example, to subvert the refutation of "All swans are white" in New Holland by reclassifying the birds in question as swimming ravens? Classification is an essential bedrock of science, but it is not unproblematic. Sound classification schemes—such as the Linnaean tree classification of biological species, or the

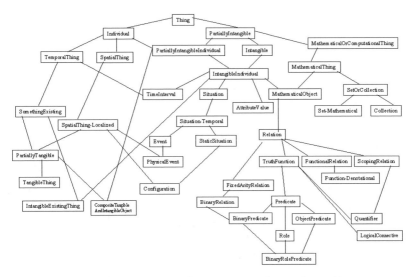

FIGURE 4.1 The Cyc Project's Upper Level Ontology diagram (© 1996 Cycorp, reprinted with permission).

mixtures, with the elements classified in the periodic table— are themselves major achievements of science. But how do we know a classification is good?

The Argentine fabulist Jorge Luis Borges imagined the "Celestial Emporium of Benevolent Knowledge," a Chinese encyclopedia that classified animals according to the scheme:

1. those that belong to the Emperor
2. embalmed ones
3. those that are trained
4. suckling pigs
5. mermaids
6. fabulous ones
7. stray dogs
8. those included in the present classification
9. those that tremble as if they were mad
10. innumerable ones
11. those drawn with a very fine camelhair brush
12. others
13. those that have just broken a flower vase,
14. those that from a long way off look like flies.[8]

59

Michel Foucault and some of his followers made fools of themselves by appearing to believe the encyclopedia was real, leading to justified complaints about the degeneracy of the postmodernist academy.[9] But the Emporium is plainly a joke, because it includes as much as possible that which would make nonsense of it as a real scheme of classification: self-reference, wildly different sizes of categories, overlapping and non-exhaustive categories, a combination of objective and human-focused principles, a mixture of the existent and the nonexistent.

A scientific classification must not make any of those mistakes. There is nothing wrong with mixing human interests into a classification when the purpose of the classification itself relates to human purposes—for example, the legal classification of animals or a classification scheme for tools or wind instruments. But mainstream science needed to extract itself from certain more primitive perspectives, or "folk" concepts.

Serious research on the question of how to classify has proved necessary in work on artificial intelligence. To enable basic inferences like:

Birds usually fly
Tweety is a robin
Therefore, Tweety (probably) flies

it is necessary to know the classification tree of the entities involved—here, that robins are a kind of bird—so that entities can "inherit" properties from further up the hierarchy.

Most helpfully, human psychology has identified in many cases a "basic" or preferred level of classification. An animal could be called "tortoiseshell," "cat," or "mammal," but humans prefer the middle level, "cat" (children naturally learn it first, natural languages give it the simplest name, subjects in time trials classify faster at that level). That agrees with scientific thinking about species in biology, but a similar phenomenon occurs in classifications of artificial objects, where the middle level of "easy chair," "chair," and "furniture" is the preferred one.[10] As in many instances, science builds on the powerful learning procedures of the untutored brain.

It is not an accident, either, that the language of the more general categories, like "mammal" and "furniture," is not "our" natural language, English, but Latin or, occasionally, Greek. That reflects the long tutelage of our language in the school of the ancients, when the technical vocabulary of English was formed by borrowings from the medieval philosophers. To appreciate the benefit received from this instruction, consider the current section in general, or the present sentence in particular, or examine adjacent portions of the text, for a variety of natural and artificial examples of philosophically influenced abstract vocabulary.[11] Science would be impossible without it.

NOT SCIENCE: (BRITISH) LAW CLASSIFIES ANIMALS

Since Roman Law times, the main legal division of animals has been into tame (which have human owners who are liable for damage they cause) and wild (*ferae naturae*). Thus sheep are easy to classify and wild lions are easy, but the behavior of some species means that whether they are wild depends on the context in which they are encountered. Bees are *ferae naturae*; when hived they become the qualified property of the person who hives them, but become *ferae naturae* again when they swarm. Parrots may become, but young unacclimatized parrots are not, "domestic animals." A performing bear is not a domestic animal, nor is a caged lion or a tame seagull used in a photographer's studio. Camels are domestic animals. So are linnets trained as decoy birds. The phrase "bird, beast or other animal, ordinarily kept in a state of confinement" includes a ferret.

A decision is also needed on the general question of what is an animal and what is not in applying laws against cruelty to animals. Cocks were not considered animals under the Cruelty to Animals (Scotland) Act 1850 but the Protection of Animals (Scotland) Act

1912 redefined "animals" to include "fowls." In the Performing Animals (Regulation) Act 1925 "animals" did not include invertebrates.

"Animal," "Domestic Animal," "Ferae Naturae" articles from John S. James, ed., *Stroud's Judicial Dictionary of Words and Phrases*, 5th ed. (1986).

A good classificatory scheme is one that finds "natural kinds" or "carves nature at the joints."[12] That is a metaphor hard to explicate fully, but it relies on the fact that the properties that objects have often come in clusters: cats have similar shapes, fur, and habits, mice have another cluster of characteristic properties, and there are no objects found with properties midway between cats and mice. The "joint" or division or large conceptual space between cats and mice is therefore natural, and the naturalness of the classification plays a role in scientific inference, in that an object with many of the properties of a natural kind can be inferred to have the rest (if it walks like a duck and quacks like a duck, it probably is a duck). The essence of classification is therefore the covariation of properties. That explains why the classification of plants makes much of obscure parts of the insides of the flower and not much of color. Color, though prominent to us, is often variable within species, whereas parts of the reproductive organs are stable indicators of species membership and hence of other inheritable traits.

The fields of numerical taxonomy and cluster analysis reduce these ideas to quantitative methods.[13] They are successful where the objects of study admit it, but are of controversial import where there are less clear divisions between kinds, as with psychiatric disorders.

The Reality of Properties, Dispositions, Laws, Causes, and Relations

The traditional "problem of universals" asks about the reality of the general categories mentioned in natural and scientific language. According to both common sense and advanced science, physical things act by virtue of the *properties* they have. A table looks as it does because of the color and shape it has. A sun attracts a planet because of the mass it has. Another table with the same color and shape (and texture and whatever other properties contribute to looks) would look the same; another body with the same mass would attract the planet in the same way. It is these repeatable properties like blue, being cubic, and having a mass of one kilogram—traditionally called "universals"—whose existence is in question. Should we say that these colors, shapes, and so on really exist, or only the things that have them?

A first reason for believing in the reality of these properties is that we only perceive objects and know *their* reality by virtue of those properties these objects have. I know the table exists because I see its color and see and feel its shape; I see and feel its color and shape because the color and shape affect my sensory organs in a particular way. But anything that has causal power, such as to affect my senses, must be a reality. Nonentities don't act.

The lawlike connections between the properties of tables and how my senses are affected is a special case of the more general laws studied by science. What is the difference between a true law of nature, like "all bodies attract one another" and a mere cosmic coincidence? Or is a law simply a summary statement of what actually happens? Surely the reason that the law supports predictions—that two new bodies would also attract—and counterfactuals—if there had been two bodies there, they would have attracted—is that there is some real connection in things between having mass and attracting other bodies. That "real connection" is a law of nature, and the aspect of things, such as mass, that it connects must also be real.[14]

Lastly, some truths are directly about universals themselves, not about the things that have them. "Orange is between red and yellow" is about *colors*, not about colored things. Much of science is an effort to understand the relations between universals, and much of experiment on objects is an attempt to find how they act in virtue of the properties and relations they have. Thus science is committed to the reality of properties.

Science also distinguishes between categorical properties (shape, for instance) and dispositional ones (solubility, brittleness). Shape is a simple property of a body, always on show. But solubility is essentially a matter of a counterfactual: *if* salt *were* put in fresh water it *would* dissolve. Dispositions can exist without being manifested, since it may happen that a soluble salt never is put in water, whereas a categorical property *just is*.[15] Naturally, there are difficult issues concerning the relation between dispositions, powers, causes, and laws of nature; perhaps some of these can be analyzed in terms of others. But it is not possible to do science entirely without them.

The notion of *cause* remains crucial to science, even though the most general physical laws do not mention causes. No physical laws or interpretations of those laws call into question such facts as that some diseases are caused by viruses and therefore that eliminating the viruses will cure or prevent the diseases. Every technological application of science requires the notion of an intervention that will effect change (or a precaution that will prevent change). That physical laws are descriptive does not undermine the notion of causality. The motions of billiard balls in interaction is described and predicted by purely descriptive laws of conservation of momentum and energy, for example. That does not in any way supersede our understanding that one ball hit another and caused it to fly off. The laws just describe the course of the causal interaction "from the outside." It is a description complete in one way but partial in another, in the same way as a complete description of a person's actions without reference to their motivations, or, as the "strong programme" sociologists of science hoped to achieve, an account of the "inscriptions" of communities of scientists without reference to the reasons they had for believing their theories.

How to know what causes what from observing what follows what is a difficult problem, at the cutting edge of cognitive science research. It will be considered in chapter 11. As to whether some other of the entities mentioned in science are real, such as forces, space, and numbers, that debate also will be left to later chapters.

Space, Time, and Continuity

Science agrees in large part with common sense on the role of space and time. Physical objects exist in space. On the medium scale that we perceive, space is three-dimensional, flat, homogeneous (the same at all points and in all directions), and continuous (indefinitely divisible as opposed to discrete or atomic). Time is one-dimensional and also, on our scale, homogeneous and continuous. The laws of nature are also invariant with respect to space and time, in the sense that particular points of space and time are not relevant to them: the laws of nature are not different before and after a certain date, nor different on the other side of the hill. Nor does the mere passage of time or change of place have an effect: if something decays over time, for example, one expects to find a positive cause other than the mere passage of time itself.

What is crucial about both space and time is the relation of *closeness* in them. Which parts are close to which others is crucial to causal processes, because they act only *locally*, that is, things affect immediately (in time) only things immediately adjacent to them (in space). Thus causal processes must propagate continuously in space and time, via such processes as the transmission of force through rigid bodies and waves. That is fortunate for our ability to predict, since to work out what will happen to an object in the near future, we need only know the causal state of things in its immediate neighborhood or in its immediate past (and not such distant things as the star-sign under which it was born). That explains the crucial role in laws of nature that is played by partial differential equations, the mathematical language for describing the propagation of local effects. We will consider these further in chapter 7.

Twentieth-century developments in relativity and quantum physics made certain, though relatively minor, qualifications to this picture. Both, indeed, were very determined applications of the idea of strictly local actions. Newtonian physics, for all its differential equations, included an exception on the large scale in its assumption that any change in gravitation would be propagated throughout the universe instantaneously—for example, if a body suddenly changed place or went out of existence, the gravitational effect of the change would be registered at once everywhere. Einstein's theory of relativity was a rearrangement of large-scale physics in reaction to the discovery that there was an absolute limit to the speed of propagation of gravitational or other influences, namely the speed of light. Relativity is thus more intuitively satisfying because it is a more strictly local theory than what it replaced. Quantum mechanics, too, is firmly based on the Schrödinger equation, a partial differential equation that describes the propagation in continuous space and time of the quantity of causal interest, a quantity that implies the probability of occurrence of various observable events.

Quantum mechanics does, it is true, claim there is something like a non-local effect, in those instances where two particles with opposite spin, for example, are emitted on divergent paths, and the later measurement of the state of one of them determines the state of the other, however distant it may be. However, as theorists insist that this phenomenon cannot be used to send a message from one place to the other, obviously this is a "causal effect" in an attenuated and technical sense.

A consequence of Einstein's changes, however, was that there is a certain unexpected connection between space, time, and motion. Einstein discovered the relativity of simultaneity, meaning that there is not a single global space and a single global time. Instead, at each point and for each "observer" (that is, for each speed at a point), there is a three-dimensional space and one-dimensional time, but the spaces and times for different observers who are moving relative to one another do not fit together perfectly to give a global space and a global time. Physicists often think of a single global "space-time" or "block universe," responsible for the different local spaces and times for any observer at any point. But global space-time can also be considered a mathematical fiction. The physics supports equally well the view that the real objects in motion "create" their own time at each point, the interactions between these many local times being constrained by the propagation of influences such as light and gravitation.

Modern physics, though having discovered that matter and energy are discrete, has not found any reason to believe that space and time are also discrete. The question is an empirical one. Discrete space and time have occasionally been proposed to account for this or that physical phenomenon, but none of these suggestions have become established. Some physicists have been unhappy with the assumption that space and time are continuous, or infinitely divisible, since the assumption requires heavy mathematical machinery while being impossible to confirm experimentally, as there are limits to how finely space and time can be measured.[16] The matter remains unresolved. All that is definitely known is that, down to the limits of observation, there is no reason to believe space and time are other than continuous.

Is space real? That is, is it a kind of stuff or ether in which objects swim, or is it just a manner of speaking about the distances between objects? By and large, the findings of science suggest a realist view of space. Science has tasks for space that include being curved (which has causal consequences for motion); supporting the propagation of gravitational, magnetic, and electrical fields; and being continuous (or not, as the case may be). Those are tough jobs for the nonexistent. An old argument of Kant's dramatizes our intuitions of the reality of space: If I take my right hand out of the part of space it currently occupies and try to fit my left hand into that part of space, I cannot do it. The part of space and its shape must be real in order to be able to perform that feat.[17]

Science has had less success in unraveling the mysteries of time. "Absolute, true, and mathematical time, of itself, and from its own nature, flows equably without relation to anything external," in Newton's magisterial phrase.[18] But it is unclear even whether any sense can be given to the notion of the "flow" of time, let alone whether there is anything equable about it, or whether there might be branching time or whether quantum mechanics has identified cases of causation backward in time. There is a certain clarity to the second law of thermodynamics, which states that disorder increases with time—for example, mixing an egg will scramble it but cannot unscramble it. The reasons for that, and whether there are possible universes in which the reverse is true, remain matters beyond our ken.[19]

Science Adjusts the "Manifest Image"

While science is largely in agreement with common sense and our grammatical categories when it comes to the basic furniture of the world, it has made, in the course of time, certain adjustments to the "manifest image"[20] of the world. In certain, though limited, respects, "folk" concepts have turned out to be inadequate.

First, a number of human-centered concepts have turned out to have no application, or at least no application that usefully bears on scientific questions. The Renaissance naturalist Aldrovandi arranges his chapter "On the Serpent in General" under headings that include: the meanings of the word, etymologies, form and description, anatomy, nature and habits, voice, diet, physiognomy, antipathy, sympathy, modes of capture, wounds caused by, remedies, epithets, prodigies, mythologies, gods to which dedicated, emblems of, proverbs, coinage, miracles, riddles, heraldic signs, dreams, statues, use in diet, use in medicine, miscellaneous uses. Paracelsus says:

Behold the Satyrion root, is it not formed like the male privy parts? No one can deny this. Accordingly magic discovered it and revealed that it can restore a man's virility and passion.... Siegwurz root is wrapped in an envelope like armour, and this is a magic sign showing that like armour it gives protection against weapons. And the Syderica bears the image and form of a snake on each of its leaves, and thus, according to magic, it gives protection against any kind of poisoning.[21]

It is argument from analogy gone mad, and it is no part of science. Human signs should be kept clear of science. Nor does science need astrology, divination, alchemy, the great chain of being, ley lines, *feng shui*, or analogies between the microcosm and the macrocosm. Those ideas could have turned out to be productive, but on the evidence, they didn't.

Science has also cast certain doubts on the truth of how we normally perceive objects. Science suggests that the "secondary qualities" such as color, taste, smell, and sound are as much in us as in the thing itself (unlike "primary qualities" such as shape, size and hardness). And it suggests that, contrary to appearances, objects are mostly empty space. It would not do to exaggerate either of these claims. Color is certainly not wholly "made up": if we perceive two surfaces as of different colors, there is a basis for that in what wavelengths of light they reflect (or sometimes differences in how adjacent surfaces are reflecting). Those are differences in the real qualities of the objects, so even if our minds add a "perceived feel" to color, we are perceiving something real by color's samenesses and differences.[22] It is the same with smell—different smells generally indicate different chemicals in the air, so smell perception is of something real. The claim that ordinary objects are "mostly empty space" also needs certain qualifications. While it is true that most of the mass of bodies is concentrated in atomic nuclei, which do indeed occupy a tiny proportion of space, the impenetrability and perceived continuity of bodies is due to the effects of electron "clouds," which are not exactly particulate and in some sense do occupy a lot of space. The space between nuclei is far from "full," but it is not entirely empty either.

LAMENT FOR THE MANIFEST IMAGE

That the glory of this world in the end is appearance leaves the world more glorious, if we feel it is a show of some fuller splendour; but the sensuous curtain is a deception and a cheat, if it hides some colourless movement of atoms, some spectral woof of impalpable abstractions, or unearthly ballet of bloodless categories. Though dragged to such conclusions, we cannot embrace them. Our principles may be true, but they are not reality. They no more *make* that Whole which commands our devotion than some shredded dissection of human tatters is that warm and breathing beauty of flesh which our hearts found delightful.

Francis Herbert Bradley, *The Principles of Logic* (1883), 501.

The concept for which scientific developments have most directly contradicted common sense is perhaps energy. There is indeed no naïve concept that is much like the scientific concept of energy, but concepts like the tendency of moving bodies to effect actions (*vis viva* in Leibniz's terminology) and also of heat are very natural. The interconvertibility of these two—and also of "potential energy"—and the conservation of total energy was certainly a remarkable discovery.

The adjustments of naïve concepts needed to cope with heat and energy are as substantial as those allegedly needed in quantum mechanics. And that was before Einstein's celebrated

$E = mc^2$, according to which *matter* and energy are also inter-convertible. That really does offend the common sense encoded in our folk categories.

Some have gone further and argued that it is not just astrology and the like that science has banished, but the mental, the ethical, and the divine. As will be argued in the last chapter, the demise of those entities has been much exaggerated.

New Concepts

Science sometimes needs new concepts. Not just new theories expressible in old concepts, like Copernicus's heliocentric theory or Harvey's theory of the circulation of the blood, but genuinely new concepts. It is not easily done, and usually not done suddenly.

Take the example of speed and acceleration. Ancient languages do not have units of speed, like miles per hour. Of course they could express "fast" and "slow," or how many days it took by horse from Rome to the Rubicon, but the concept of any numerical measurement of speed, or of one speed being twice another, was missing. The Latin word *per*, which we use to form units of rates (miles per hours, grams per cubic centimeter, etc.), is only used in that sense since late medieval times, originally with money (such as costs in pence per measure or per head). The idea that a body has a measurable speed at one moment that may or may not equal its speed at later moments needs both some experience with the measurement of small time intervals—impossible before the seventeenth century—and careful conceptual analysis. More and harder conceptual analysis is then needed to distinguish speed from acceleration and to realize that, in principle, acceleration too is a measurable quantity, with units like miles per hour, which itself may or may not change over time. These distinctions were made by the Merton School of philosophers at Oxford in the fourteenth century and their associates in France. They understood, for example, how acceleration, speed, and distance traveled determine one another: for example, if a body starts from rest and has a constant acceleration, its speed increases linearly with time and its distance traveled increases quadratically with time—thus in times 1, 2, 3, 4 … seconds it will have traveled distances from the start proportional to 1, 4, 9, 16….[23]

These theorists missed an important discovery in not realizing that heavy bodies dropped in air *do* have constant acceleration. As this acceleration is quite large—32 feet per second per second, thus with a barely perceptible delay between dropping a rock and pain in the foot—it is hard to establish this without slow-motion photography. One of Galileo's most remarkable achievements was to argue convincingly for the constant acceleration of falling heavy bodies. He first showed, with a cunning mathematical argument, that the most natural alternative theory—that the speed of a body was proportional to the *distance* fallen (as opposed to the time taken)—was actually impossible. Then he experimented with bodies rolling slowly down inclined planes, and used a continuity argument to show that falling bodies would behave similarly, with constant acceleration. His achievements were extraordinary, but the actual conceptual machinery he needed had been supplied for him by thinkers two and a half centuries earlier.

It is no accident that the example is mathematical. Mathematics is the source of most of the genuinely new concepts in science. These include probability (of different kinds, to be described in chapter 10); the calculus of Newton and Leibniz that gave precise meaning to the quantifying of continuous processes; the field theory of Faraday and Maxwell; Riemann's geometry, which supplied the language of general relativity; exponential growth; chaotic dynamics in population growth; the double helix of DNA; feedback in engineering; information-processing models of perception; the statistical design of experiments in agriculture and psychology; and stock volatility in option pricing.

In chapter 7, we will look at how knowledge of mathematics and its applications works.

CHAPTER 5

The Physical Sciences

"The proper study of mankind is man," Alexander Pope says. From a poet, an acceptable sentiment—almost a requirement of the job description. Humans are naturally and rightly of special interest to themselves. But let us take a break from species-narcissism now and again. There are many other things in the universe, living, non-living, and abstract, all with their alien ways of working. It is human to want to understand their activities. "Happy he who knows the reasons for things," Virgil says (though a poet), possibly referring to Lucretius and his pioneering attempt to demystify the world on scientific principles, *De rerum natura*. Those things are interesting exactly because they are strange to us, not subject to our wills nor open to our negotiations. As early man quickly realized, the arrangements of the stars are unaffected not only by the rise and fall of princes but by any earthly cataclysm. They do not need to be "constructed as other," as the postmodernists say. They *are* other.

The natural sciences come in two classes. In one class are the sciences of mid-range things and stuff—chemistry, zoology, agricultural science, anatomy, neuroscience, ornithology, oceanography, geology, solid state physics, solar system astronomy, and so on. They deal with things of moderate size and complexity, accessible to observation and experiment. Knowledge in them tends to be cumulative, as better observational techniques allow theory to extend its range gradually. The other class has a membership of one: fundamental physics. Knowledge there comes up against the problems of the very small and the very large, and the connection between theory (abstruse and mathematical) and observation (expensive) is stretched to its limit.

Extremities of scale are not the only two dragons that inhabit the regions beyond the known conceptual world. The other two are the past and complexity, which, as we will see, inhibit knowledge not only of very remote events like the Big Bang but also of complex processes whose pasts are not easily accessible, such as the origin and evolution of life and the system of global climate.

REGIONAL OCEANOGRAPHY: SCIENCE OF THE PARTICULAR

- By far the world's largest ocean current is the Antarctic Circumpolar Current, driven by westerly winds around 60°S where the earth's surface is ocean all the way round; an entire ocean is moving east with a flow well over 100 times that of all the world's rivers combined; it preserves the Antarctic ice cap
- Most ocean currents are, like the ACC, wind-driven, but a few are not; the North Atlantic Current that extends the Gulf Stream and gives Western Europe a climate unlike Siberia's is a potentially unstable thermohaline circulation, kept in motion by subtle differences in temperature and salinity
- A massive northerly current in the Atlantic transports heat across the equator, making the northern hemisphere warmer than the southern (at equal latitudes)

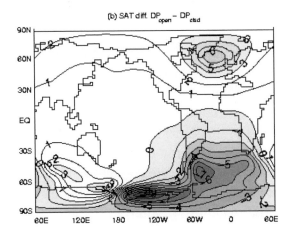

(b) SAT diff. $DP_{open} - DP_{clsd}$

Effect on world temperatures of the creation of the Antarctic Circumpolar Current 30 million years ago, when Antarctica and South America separated (blue: colder, red: warmer). From Willem P. Sijp and Matthew H. England's "Effect of the Drake Passage Throughflow on Global Climate," *Journal of Physical Oceanography* 34 (2004): 1254–1266, courtesy of the authors.

- Circulation along the ocean floor has different and slower patterns; cold salty water sinks to the bottom of the Weddell Sea off Antarctica, and can reach the deepest parts of the ocean in the trenches of the North Pacific over a thousand years later
- The Mediterranean Sea evaporates more water than it receives from rivers, thus has to receive normal seawater through the Strait of Gibraltar and give out (at a lower level) considerably saltier water
- The Black Sea, on the other hand, is much fresher and, in its deeper waters, dead (anoxic), while the Baltic is close to fresh and only receives occasional influxes of salt water
- The El Niño-Southern Oscillation is a disturbance of the world's currents and weather, centered on the south-eastern Pacific, which recurs irregularly every 2–7 years and induces droughts in Australia and South America; its cause is unknown

Summarized from M. Tomczak and J. S. Godfrey, *Regional Oceanography: An Introduction* (Oxford, England: Pergamon, 1994).

Physics

Physics is special. It is a science that applies to all material things, since its subject matter is just the properties of material things as such. It is the science of the largest, smallest, and oldest things. Its findings underpin the reliability of all engineering. And it is rightly regarded as the foundation of all the other sciences, in that other sciences can in a certain sense be reduced to physics. That is a claim about the facts rather than about our knowledge. Leaving aside controversial possibilities such as gods, minds, and free will, it appears that physical laws determine chemistry, while (bio)chemistry determines the properties of living things (along with, of course, physics itself and geometry, such as the weight and shape of elephant legs). At no point is there a need to invoke fundamentally new explanatory principles such as "vital forces," even though the behavior of large assemblages (of atoms, of cells) cannot always practically be predicted from the behavior of the components. Whether the biological properties of the brain wholly determine the process of thinking is not so clear; we examine consciousness in the last chapter.

REDUCTION: THE COLORS OF TRANSITION METAL COMPOUNDS

The compounds of elements at the left of the Periodic Table are of generally uninteresting color: common salt (sodium chloride), chalk (calcium carbonate or sulfate, depending which kind is meant), potash and baking soda are all white or off-white. Then as one moves to the "transition metals" in the middle of the table, their compounds break out into a spray of bright colors: rust and blood (from iron), deep purple condy's crystals (potassium permanganate), copper sulfate (blue), verdigris (copper carbonate), viridian (chromium oxide), sapphires and rubies (aluminium oxide with iron, titanium and chromium impurities), emerald (beryl with traces of chromium), turquoise (a phosphate of copper and aluminium), vermilion (mercuric sulfide), cobalt blue, chrome yellow.

Like any physico-chemical phenomenon, the explanation is in terms of the solutions of the Schrödinger equation of quantum mechanics, which determine the electron orbitals around the different atoms and their compounds. To simplify greatly, it so happens that for transition metal compounds, the differences in energy levels in the orbitals are such that photons in the visible-light range can cause a change in the levels by being absorbed. Thus when white light is shone on such a compound, certain characteristic frequencies are removed from it, resulting in the reflected light being seen as colored.

A complex but still simplified story at www.chemguide.co.uk/inorganic/complexions/colour.html.

Physics also has certain advantages when it comes to knowledge. There are no ethical problems about experimenting on lifeless matter—one may metaphorically "torture" Nature to reveal its secrets, as Francis Bacon put it, but matter does not literally suffer. And with certain exceptions, such as in quantum mechanics, the response of matter to measurement and experiment is extremely consistent.

Just as well, because some of the objects of physics, such as the very large, very small, and very old ones, are hard to get at.

Precision Measurement: Source of Physical Theory

Physics and associated natural sciences admit a degree of precision—and a corresponding lack of need for much statistical analysis—rarely found in other, more furry, sciences. Much of what physics is trying to study, plain matter, is well-behaved when it comes to variability. Electrons all have the exactly the same charge. The gravitational force of two masses fits Newton's law of attraction to the utmost degree of exactitude measurable. That allows precision measurements to be a tool for discovery in those sciences in ways impossible elsewhere.

An example is the discovery of argon. Mendeleev's Periodic Table of the Elements originally lacked the entire right-hand column, containing the inert or "rare" or "noble" gases (helium, argon, neon, and others). The reason they were hard to discover was exactly that they are inert—their outer electron orbitals are full, so they very rarely enter into chemical reactions with other elements. They are also colorless and odorless. But they are not exactly rare—apart from the helium in the sun (discovered by the analysis of spectra) about 1 percent of the earth's atmosphere consists of inert gases, mostly argon. That fact was discovered through close attention to precision measurements.

Late in the nineteenth century, the physicist Lord Rayleigh made precise measurements of the density of nitrogen. Nitrogen, known by then to be the main constituent of the atmosphere, can be obtained in two ways: by removing from air all the more reactive constituents such as oxygen, water vapor, and carbon dioxide, or by decomposing ammonia, a compound of nitrogen. It was expected that the densities of the nitrogen obtained by the two methods should be exactly the same; it was understood by that time that the elements were very well-behaved in having no variability of properties in different pure samples. But the density of the sample of nitrogen obtained from air was slightly higher. The discrepancy is only about one part in a thousand, which in almost any science except physics would be well below the level of experimental variability. But the material with which physics deals is extremely consistent in its behavior. With enough repetition of precision measurements, it was established that the discrepancy was real. After ruling out some hypotheses such as a heavier form of nitrogen molecule (similar to ozone for oxygen), Rayleigh was able to remove the nitrogen itself from the air sample by chemical reaction and find that there was a small residue of an even more unreactive gas. It proved to have consistent physical properties and was identified as the element argon (with small admixtures of the other inert gases).[1]

THE TWO CULTURES: *YES MINISTER*

A chemical plant is to be built to fulfill a huge contract to make metadioxin, a harmless chemical which unfortunately sounds like the dangerous dioxin. Minister Jim Hacker discusses the problem with the member in whose constituency the plant will be built, and Sir Humphrey and Bernard. Being classically trained public servants, Humphrey and Bernard become distracted by the question of the lack of an ablative in Greek. They are recalled to the matter at hand:

> Sir Humphrey: Metadioxin is an inert compound of Dioxin.
> Local member: What?
> Jim: Yes I think I follow that Humphrey but, er, could you, explain it a little more clearly.
> Sir Humphrey: In what sense Minister?
> Local member: What does inert mean?
> Sir Humphrey: Well it means it's not … ert.
> Bernard: Wouldn't ert a fly.

"The Greasy Pole," *Yes Minister,* BBC TV series

Thought Experiments

Physics is the natural home of one of the strangest and least understood methods of knowing in science: thought experiments. The idea of doing an experiment in one's head seems inherently contradictory: surely the point of *experiment* is to pose a question to the real world, the very opposite to letting one's imagination run riot? Yet a number of classical thought experiments, as purely "in the mind" as Descartes's examination of his own doubts, have played crucial roles in the development of physics, from Galileo's ship to Newton's cannonball to Einstein's moving clocks to Schrödinger's cat. How do they work? How can they possibly work?

Galileo's *Dialogue Concerning the Two Chief World Systems* defends the Copernican theory that the earth circles the sun. The theory includes the initially implausible hypothesis that the earth, huge as it is, spins on its axis once a day. The implausibility comes from the fact that this motion, immensely fast as it must be at the earth's surface, is completely unfelt by us. To take just one example of what we ought to see: surely a ball dropped straight down from a tower should fall to the west of the tower, which has moved a great distance east in the time of fall?

The figure who represents Galileo in the dialogue replies with a thought experiment to illustrate the relativity of motion. It is rich in circumstantial yet relevant detail:

Shut yourself up with some friend in the main cabin below decks on some large ship, and have with you there some flies, butterflies, and other small flying animals. Have a large bowl of water with some fish in it; hang up a bottle that empties drop by drop into a wide vessel beneath it. With the ship standing still, observe carefully how the little animals fly with equal speed to all sides of the cabin. The fish swim indifferently in all directions; the drops fall into the vessel beneath; and, in throwing something to your friend, you need throw it no more strongly in one direction than another, the distances being equal; jumping with your feet together, you pass equal spaces in every direction. When you have observed all these things carefully (though doubtless when the ship is standing still everything must happen in this way), have the ship proceed with any speed you like, so long as the motion is uniform and not fluctuating this way and that. You will discover not the least change in all the effects named, nor could you tell from any of them whether the ship was moving or standing still. In jumping, you will pass on the floor the same spaces as before.... Finally the butterflies and flies will continue their flights indifferently toward every side, nor will it ever happen that they are concentrated toward the stern, as if tired out from keeping up with the course of the ship, from which they will have been separated during long intervals by keeping themselves in the air. And if smoke is made by burning some incense, it will be seen going up in the form of a little cloud, remaining still and moving no more toward one side than the other. The cause of all these correspondences of effects is the fact that the ship's motion is common to all the things contained in it, and to the air also.[2]

The question is, why do we agree with Galileo? How do we know that the results would be as he says if we went on a ship? We use our imagination, but how does the imagination know? How, that is, does the imagination acquire the structure that mimics the world, and that allows it to be used as a bridge between real experiments and what would happen in counterfactual circumstances?

One way or another, the imagination must be structured by the flux of ordinary experience. An indication of how it happens can be had by recalling Stevin's Wreath of Spheres diagram of 1586, a thought experiment of the same kind as Galileo's.

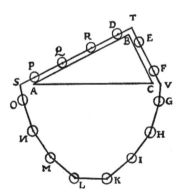

FIGURE 5.1   Stevin's Wreath of Spheres diagram of 1586.

The point of the diagram is that we can see [*sic*] that the string of spheres does not move of its own accord either clockwise or counterclockwise. It is in equilibrium as it is. Therefore, the spheres on the gentle slope, though many, balance the spheres on the steep slope, though few (as predicted by what we now call the resolution of forces).

That the spheres do not revolve in either direction, but hang in equilibrium, is a deliverance of the imagination, but not a logical truth. It must, therefore, be a distillation of experience. It follows that the process of using the imagination to learn the results of thought experiments is a kind of memory—not a snapshot or "episodic" kind of memory, but a generalizing kind of memory, which remembers the sort of thing that happened in the same sort of circumstances. As Galileo realizes, this is close to Plato's theory that knowledge is reminiscence (from a past life).[3]

This raises the problem of whether the person in the street, or the physics student in the lab, can correctly remember such things without the Socratic questioning of Galileo to assist him. Psychologists have investigated, and the answer is … yes and no. Consistent patterns of expectation about motion are found on eliciting subjects' "intuitive physics" or "naive physics" by asking them to imagine what motion would occur in various circumstances. Some subjects give the correct answers, but a high proportion, even those educated in physics, persist in making such mistakes as expecting curved motion to continue in a curve when released.[4] So thought experiments are useful—presumably accurate enough to give us while driving the means of anticipating the traffic a few seconds ahead—but it might not be wise just yet to close down the physics laboratories where they measure the real behaviors of masses.

Outside physics, there are few thought experiments. There seem to be none in chemistry (though there may be some in cookery, in the sense of imagining with some degree of correctness the likely outcome of recipes with non-actual combinations and proportions of ingredients). The reasons are unclear. There are a few in biology, perhaps too many in evolutionary theory, such as Darwin's imagining the outcome of competition between giraffes with different lengths of neck.[5]

Fundamental Physics and the Edge of Knowledge

Those who approach physics with a philosophical interest usually make a beeline for quantum mechanics (QM). That is natural because the results of QM are the foundation for most of what is known about the physics and chemistry of the small and medium scale. QM is wide in range, extremely well-confirmed, and gives exactly accurate predictions of experimental results (to the extent that is probably the most successful scientific theory ever). But it is also an unfortunate choice for philosophical attention, because of its long-running and intractable problems with interpretation. It has managed to move beyond early problems caused by its founders in the 1920s being immersed in the seething cauldron of German philosophical speculation and expressing themselves in phrases suggestive of "reality dependent on The Observer."[6] Modern interpretations are more realist, but the conflict between them is such that any rational person who is not a highly trained expert in the field should simply tell the QM industry to go away and report back to the rest of the intellectual world when they have sorted it out.

Given that QM is a highly successful theory, and that it is more than simply a technique for deriving experimental results, it is fair to ask that the experts should explain what is known, and distinguish it from what lies beyond the boundary of the known in the realm of speculative interpretation. It is extraordinarily hard to find a clear answer, or any answer, to that very reasonable request. Let us attempt the basics of that task.

Any "system"—for example, a free electron moving through space, or a hydrogen atom—has a "wave function" associated with it that is supposed to describe the total truth about it. To head off trouble later, it is wise to say immediately that the use of a count-noun like "electron" does not imply that the entity so named must be particulate or exactly localized. The wave function is a function of space and time—ordinary continuous space and time—meaning that it gives the "strength" of the system at any point in space at any time. It is thus similar to the function that describes a wave in water by giving its height at any point, at any time. There is no problem about interpreting the wave function realistically, as describing a definite physical quantity of the system. It is however not directly observable. The wave function also evolves in time completely deterministically (if the system is undisturbed), according to the Schrödinger equation (which thus plays a role similar to the equation that describes how water waves evolve).[7]

Problems arise over two matters: What *is* the property of the system that the wave function describes? And what is its relation to what can be observed, that is, to experimental results? Relevant to both these questions, but insufficient to answer them fully, is the fact that the wave function contains enough information to calculate the probabilities of the various possible outcomes of experiments on the system. For example, the wave function of a free electron will say that at any given time, if one observed the position of the electron, it would be very likely to be in one region and unlikely to be in others. The observation of an electron's position reveals a particle at a particular point (that is, a discrete event). "Probability" here means factual or stochastic probability as will be described in chapter 10—it is a physical property or tendency of the system and experimental setup, similar to a bias in dice, that means that if the identical experiment is repeated many times, the outcomes will tend to be in proportion to the probabilities.

So much is free of doubtful interpretations and—in combination with the great amount of knowledge of the wave functions of various systems and the abilities to calculate with them—delivers the vast array of successful results of quantum mechanics. Where one crosses the line into interpretation is if one falls to the temptation to say that before the observation, the electron was a particle at some position, and the wave function gives the probability of its being at that position. The established theory does not say any of those things—neither that there *is* a particle with (or without) a position, nor that the wave function gives a probability of anything to do with such a particle. The wave function is continuous and the reality it describes is not particulate. So let us keep to what is known: that the wave function describes the state of the system, and that the state of the system determines the frequency of various observational outcomes.

That quarantines the mysterious aspects of the central part of QM to one small (but crucial) phenomenon: the sudden transition of the system from the continuous reality described by the wave function to the particulate reality of, for example, an electron registering on a screen in an experiment. It is generally agreed that in normal cases this "collapse of the wave function" has to do with the interaction between the submicroscopic system and macroscopic reality, which may, but need not, be a measuring apparatus. It thus has no relation to "observation" in the sense of an activity of human minds.[8] Beyond that, it is not safe to take a stand. As academics love to say, "Much further research is needed."

Subatomic or nuclear physics ventures even further out into the unknown. Considering that it is hard enough to deal with whole atoms and that the nucleus is only around one-hundred-thousandth the size of an atom, it is surprising that anything is known about it at all. But there is a good deal known about the size, structure, and stability of nuclei and the products of their breakup. The competing theories attempting to explain those facts are best described as colorful in their mutual inconsistency.

Astrophysics and the Big Picture
On the large scale, it is equally hard to extract from physicists' writings where the boundary lies between established knowledge

and speculation. The solar system is well-known (except for the interiors of the sun and planets). Its history is less firmly established, but there are clearly leading theories on the condensation of the solar system out of a cloud of matter and the origin of the moon from a collision between the earth and something big. The basic structure of the Milky Way and other galaxies is also known. There are millions of observed galaxies per square degree of sky, containing billions of stars each, all receding from one another in a space very close to flat and at a rate consistent with a Big Bang 14 billion years ago.[9]

SOME PEOPLE JUST DON'T WANT TO KNOW

His [Sherlock Holmes'] ignorance was as remarkable as his knowledge.... My surprise reached a climax, however, when I found out incidentally that he was ignorant of the Copernican Theory and of the composition of the Solar System. That any civilized human being in this nineteenth century should not be aware that the Earth travels around the Sun appeared to me to be such an extraordinary fact I could hardly realize it.

"You appear astonished," he said smiling at my expression of surprise. "Now that I do know it, I shall do my best to forget it."

"But the Solar System!" I protested....

"What the duce is it to me?" he interrupted impatiently: "you say that we go around the Sun. If we went around the moon it would not make a pennyworth of difference to me or my work."

Dr. Watson on Sherlock Holmes's demand for vocational relevance in knowledge from *A Study in Scarlet* by Arthur Conan Doyle.

That is a lot of stuff out there. But not, as it turns out, enough stuff to make the physics work, and it is at this point that established knowledge comes to an end. According to Einstein's General Relativity, gravity is the dominant force on large scales and so the large-scale structure of the universe is mainly determined by the amount of matter in various places. There is very much less observed matter than is needed to explain the gravitational interactions seen between galaxy clusters—only about 4 percent of what is needed. So it is postulated that "dark matter" and "dark energy" make up the deficit. The nature and properties of dark matter, other than gravitational, are unknown, as it does not seem to interact, other than gravitationally, with normal matter.[10] That is, it is purely a postulate to make up a deficit in the theory. That is reasonable speculation, but it is not knowledge.

CHAPTER 6

Biology and Cognition

The circulation of the blood, one of the most basic facts about the human body, was not discovered until the seventeenth century. That illustrates one of the problems about knowledge in biology. A living thing is a very complex system, and if it is cut open to examine how it works, it may not be a living system anymore.

There is a vast amount to be known in biology. A vast amount *is* known, partly because some of it is easy to observe—the whiteness of swans, for example—and partly because of the natural readiness of the human race to invest heavily in medical research. But there is an enormous amount yet to be known, even in matters of principle. Ignorance is close to total on the origin of life, or whether it is likely to be created in a test tube or is likely to have appeared on another planet. With certain other important topics, such as information processing in the brain, differentiation in the embryo, the dynamics of ecosystems, and the evolution of species, there is a good deal of "in principle" understanding but often not much hope of modeling, predicting, or controlling any particular phenomenon. And we have as yet little idea of the prospect of making ourselves immortal.

PERMANENTLY ESTABLISHED KNOWLEDGE IN BIOLOGY

I do not believe that it will ever be shown that the blood of animals does not circulate; that anthrax is not caused by a bacterium; that proteins are not chains of amino acids. Human beings may indeed make mistakes, but I see no merit in the idea that they can make nothing but mistakes.

Henry Harris (Regius Professor of Medicine at Oxford), "Rationality in Science," in *Scientific Explanation: Papers Based on Herbert Spencer Lectures Given in the University of Oxford,* ed. A. F. Heath (Oxford: Clarendon, 1981), 36–52.

Some biological knowledge demands hard work on the ground but logically is not very problematic: observing the differences between species, dissecting a lot of bodies to create the vast corpus of knowledge in the fortieth edition of *Gray's Anatomy*, carefully interpreting what is observed through microscopes and in MRI images, strictly applying statistical tests to determine whether new pharmaceuticals work. The confidence one has in what one's dentist forecasts is founded on that great array of work, one of the supreme achievements of the human race.

A SUDDEN ACCESSION OF KNOWLEDGE:
VAN LEEUWENHOEK'S NEW MICROSCOPIC WORLD
IN A DROP OF WATER

Delft, September 7, 1674

Passing just lately over this lake ... I took a little of it in a glass phial; and examining this water next day, I found floating therein divers earthy particles, and some green streaks, spirally wound serpent-wise, and orderly arranged, after the manner of the copper or tin worms, which distillers use to cool their liquors as they distil over. The whole circumference of each of these streaks was about the thickness of a hair of one's head ... all consisted of very small green globules joined together: and there were very many small green globules as well. Among these there were, besides, very many little animalcules, whereof some were roundish, while others, a bit bigger, consisted of an oval. On these last I saw two little legs near the head, and two little fins at the hindmost part of the body....

Clifford Dobell, *Antony van Leeuwenhoek and His "Little Animals"* (London, 1932), 110.

For all that, there are strange gaps in science's knowledge of how the parts of the body fit together to make a working whole. Those gaps mean that while surgery, for example, which deals with corrections to local bodily malfunctions, is an advanced science, diagnosis of whole-body complaints is a weak point of medicine. Feelings of "general malaise" or "failure to thrive" or chronic exhaustion often present problems for diagnosis even after all the necessary and unnecessary pathology tests have been done, and a second opinion may be a good idea. Resort to alternative medicine can sometimes be a rational course of action even if the accompanying theorizing is obvious rubbish.

The Sleep Mystery

The way in which biological complexity permits an accumulation of established knowledge intertwined with fundamental lack of understanding is illustrated by research on an old question where there is a mystery in plain sight. What is sleep for? Animals spend a large proportion of their time sleeping, they need sleep desperately and they function badly without it, in ways everyone is very familiar with after a bad night. But the answer "We need sleep to have a rest" is fundamentally uninformative. The parallel between animals and machines and between brains and computers is unhelpful, since machines and computers do not need rest—generally, they are better off operating all the time except when there is something wrong with them. And even if there is some reason why animals benefit from staying quiet to recuperate or "recharge," it fails to explain why they do something so dangerous as to turn off awareness of environmental input.

Probably the leading hypothesis on the purpose of sleep is that it has something to do with the laying down and organization of memories. The theory has initial plausibility and some empirical support, but has proved immensely difficult to establish convincingly. There is a kind of logical illusion that is characteristic of research on complex working systems such as biological ones. The natural method of experimental approach is to see what happens when some part or process is interrupted or damaged and so to draw conclusions about the causes of normal functioning. Certainly, it is easy to pile up experimental facts on the degradation of memory if humans or animals are deprived of sleep. But that suffers from a logical problem: if one removes a part from a television and the picture turns into diagonal stripes, it does not follow that one has found the horizontalizer. Complex systems work in a more holistic manner: the malfunctioning of a part may have many different effects because of its place in many causal pathways. Lack of sleep causes many cognitive problems, but that tells us little about the essential purposes of sleep. The fact that there is little idea what, on the memory hypothesis, might be the different purposes of REM (dreaming) and non-REM (deep) sleep shows how little progress the hypothesis has made. Recent work with brain imaging that allows direct observations on how sleep affects the plasticity of neurons tends to support the memory hypothesis, but still gives little insight into how sleep actually does anything for memory and learning. A 2006 review of the evidence concludes that "despite a steady accumulation of positive findings over the past decade, the precise role of sleep in memory and brain plasticity remains elusive."[1]

At the same time, and as is typical of biology, the variety of species has produced a diversity of sleep phenomena so complex as to give the impression of being designed specifically to confuse investigators trying to understand it. Small mammals like rats sleep a lot while elephants don't—but then lions sleep quite a lot, too. Rats can die from lack of sleep but are unusual in so doing. REM sleep takes up a high proportion of sleep time in babies and less in later life, possibly suggesting it is involved in processing novel experiences—but it is the other way round in dolphins. The mammal with the highest known amount of REM sleep is the platypus.[2]

Not all biological problems are as hard to solve as that of sleep. Even so, there are recurrent difficulties in understanding the overall functioning of organisms. Of the two obstacles to knowledge especially pertinent to biology, the variability between cases and the complexity of organisms, the first has proved more tractable. We will see in chapter 10 how statistical methods have made progress in drawing sound conclusions despite variability between individual cases. In chapter 13, we will examine complexity again, showing how it creates immense difficulties in firmly establishing the theory of evolution.

PALEY'S *EVIDENCES:* MACHINE AND BIOLOGICAL COMPLEXITY COMPARED

In crossing a heath, suppose I pitched my foot against a *stone*, and were asked how the stone came to be there; I might possibly answer, that, for any thing I knew to the contrary, it had lain there for ever: nor would it perhaps be very easy to show the absurdity of this answer. But suppose I had found a *watch* upon the ground, and it should be inquired how the watch happened to be in that place; I should hardly think of the answer which I had before given, that, for any thing I knew, the watch might have always been there. Yet why should not this answer serve for the watch as well as for the stone? Why is it not as admissible in the second case, as in the first? For this reason, and for no other, viz. that, when we come to inspect the watch, we perceive (what we could not discover in the stone) that its several parts are framed and put together for a purpose, e.g. that they are so formed and adjusted as to produce motion, and that motion so regulated as to point out the hour of the day; that, if the different parts had been differently shaped from what they are, of a different size from what they are, or placed after any other manner, or in any other order, than that in which they are placed, either no motion at all would have been carried on in the machine, or none which would have answered the use that is now served by it. To reckon up a few of the plainest of these parts, and of their offices, all tending to one result: We see a cylindrical box containing a coiled elastic spring, which, by its endeavour to relax itself, turns round the box. We next observe a flexible chain (artificially wrought for the sake of flexure), communicating the action of the spring from the box to the fusee. We then find a series of wheels, the teeth of which catch in, and apply to, each other, conducting the motion from the fusee to the balance, and from the balance to the pointer; and at the same time ... that over the face of the watch there is placed a glass, a material employed in no other part of the work, but in the room of which, if there had been any other than a transparent substance, the hour could not be seen without opening the case. This mechanism being observed ... the inference, we think, is inevitable, that the watch must have had a maker ... an artificer or artificers who formed it for the purpose which we find it actually to answer; who comprehended its construction, and designed its use.

I. Nor would it, I apprehend, weaken the conclusion, that we had never seen a watch made; that we had never known an artist capable of making one; that we were altogether incapable of executing such a piece of workmanship ourselves, or of understanding in what manner it was performed....

II. Neither, secondly, would it invalidate our conclusion, that the watch sometimes went wrong, or that it seldom went exactly right....

IV. Nor, fourthly, would any man in his senses think the existence of the watch, with its various machinery, accounted for, by being told that it was one out of possible combinations of material forms....

V. Nor, fifthly, would it yield his inquiry more satisfaction to be answered, that there existed in things a principle of order, which had

disposed the parts of the watch into their present form and situation ... every indication of contrivance, every manifestation of design,
which existed in the watch, exists in the works of nature; with the difference, on the side of nature, of being greater and more, and that in
a degree which exceeds all computation. I mean that the contrivances
of nature surpass the contrivances of art, in the complexity, subtility,
and curiosity of the mechanism; and still more, if possible, do they
go beyond them in number and variety; yet, in a multitude of cases,
are not less evidently mechanical, not less evidently contrivances, not
less evidently accommodated to their end, or suited to their office,
than are the most perfect productions of human ingenuity.

William Paley, *Natural Theology: Or, Evidences of the Existence and Attributes of the
Deity,* 12th ed. (1809), 1–18.

Cognitive Science

The science of the mind is of course one of the most exciting
of the sciences. It is about *us*, and about the most distinctive
part of our selves. And it is knowledge about knowledge. From
the point of view of scientific knowledge, cognitive psychology
promises to reveal the secrets of how we know—how we know
in science as in everything else.

On the other hand, the very fact that in cognitive science
knowledge is trying to study itself threatens some special problems with self-referential illusions. Can we "step outside ourselves" and make correct inferences about what is inside us from
looking at external behavior (of ourselves or others)? If a computer performed similarly "adaptive" or intelligent behavior,
what would we infer about its insides?

As a coherent science, the serious development of psychology
is only recent. Older traditions like medieval faculty psychology,
introspective psychology, psychoanalysis, and various experiments like Pavlov's on dogs produced many disjointed insights
but no overall view of the field. By the 1950s, the more scientific end of psychology was dominated by a simplistic theory,

behaviorism. According to that view, learning and action could be understood on the model of stimulus-response and conditioning, without inquiry into any hidden entities there might be inside the brain or mind.

In the same decade, computers appeared and were immediately touted as "electronic brains." The computer model of the mind was not exactly behaviorist, in that there was conceived to be something inside, namely software, driving behavior. But the model of intelligence in early AI was still a very simple one, based on calculation and search (through spaces of possibilities, such as the possible moves in a game). Alan Turing proposed the "Turing test": that a computer should be regarded as thinking if it could answer questions in such as way as to fool an observer into believing it was a human doing the answering. Turing had a reasonable sense of the difficulties that would involve—he offers these few lines of specimen dialogue:

Question: Please write me a sonnet on the subject of the Forth Bridge.
Answer: Count me out on this one. I never could write poetry.
Question: Add 34957 to 70764
Answer: (Pause about 30 seconds and then give as answer) 105621.
Question: Do you play chess?
Answer: Yes.
Question: I have K at my K1, and no other pieces. You have only K at K6 and R at R1. It is your move. What do you play?
Answer: (After a pause of 15 seconds) R-R8 mate.[3]

(The answer to the addition is wrong, with a mistake that is typical of a human but not of a machine). Turing predicted computers would pass the test within 50 years. More optimistic colleagues reduced the timescale. Allen Newell and Herbert Simon, early leaders of the Artificial Intelligence project, predicted in 1957:

1. That within ten years a digital computer will be the world's chess champion, unless the rules bar it from competition.
2. That within ten years a digital computer will discover and prove an important new mathematical theorem.
3. That within ten years a digital computer will write music that will be accepted by critics as possessing considerable aesthetic value.
4. That within ten years theories in psychology will take the form of computer programs, or of qualitative statements about the characteristics of computer programs.[4]

Everyone praised the preparedness of the AI gurus to make such falsifiable predictions, although those who might otherwise have won the grants that flowed torrentially into AI until the early 1970s were more envious than admiring.

The predictions were falsified comprehensively, and with them the model of the mind on which they were premised. Fifty years later, we are still waiting. The computer revolution has produced amazing advances undreamed of in 1957, from the World Wide Web to spectacular increases in raw computing power that make a computer of that date look like an abacus, but the predicted advances in AI have gone almost nowhere. The only one of Newell and Simon's predictions to be realized, eventually, was that a computer would defeat the human world chess champion. And even that was a Pyrrhic victory, since chess-playing computers do not play chess *like a human*. Instead they search through billions of moves per second as a substitute for whatever it is that human players do, leaving it as much of a mystery as ever what that might be. Nor is there a computer that is close to passing the Turing test.

Inference to What Is in the Mind

If an artificial mind cannot be produced, there is no choice but to infer the machinery and workings of real ones by observing what they can do.

There has been a long-standing hope that it would prove possible to connect psychology with brain science, finding the physical basis of thought by such means as neural imaging, the actions of individual neurons, brain lesions, the effects of drugs, and the like. The large sum of knowledge gained from those experiments has proved disappointing, so far, as a means of understanding how the brain actually thinks. As a former enthusiast for brain imaging explains, it produces facts that are not quite to the point or at the right scale:

> Although technologies to directly represent brain function are exciting and visually compelling, they still only provide a limited amount of detail. Imagine trying to figure out what is happening inside the Empire State Building by watching the lights go on and off. You might get some idea as to the internal function, but not much. Now imagine that you can change the pattern of lights by setting off sirens outside, directing heat at the building or sending somebody inside with a chocolate cake. You might, through careful observation, be able to work out those areas of the building that have good sound-proofing or air conditioning and where the kitchens are located.[5]

Attempts to find out what is happening inside the brain by making it work wrongly, through lesions or psychoactive drugs, face the same logical problems as mentioned about with research into sleep deprivation: the more complex a system is, the harder it is to find the purpose of a part by observing the system when that part is not working.

Psychologists, as opposed to neuroscientists, have proceeded on a different and more hopeful plan. They prefer to make inferences from observed behavior to the mental structures necessary to create it. The logic is similar to making inferences from what a calculator can do to the capabilities of the software inside— for example, if pressing <number>+<number>= on a calculator reliably causes the display of the sum of the two numbers, it may be inferred that the calculator implements addition, even if nothing is known about its wiring. By these methods cognitive psychology has made considerable progress on understanding what it is that minds do inside. The classic 1971 experiment by Roger Shepard and Jacqueline Metzler on mental rotations illustrates how such inference works. Subjects were shown pairs of pictures of 3D "arms" such as the following:

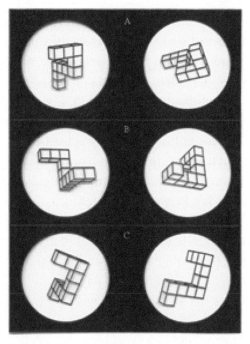

FIGURE 6.1   From Roger N. Shepard and Jacqueline Metzler's "Mental Rotation of Three-Dimensional Objects," *Science* 171 (1971): 701–3, reprinted with permission from AAS.

In some pairs, one shape may be gotten from the other by rotation while in other pairs that is not possible: they are enantiomorphs like the left and right hand that cannot be rotated to fit in the same space. Subjects are asked to state whether one is a rotation of the other and response times are recorded. It is found that, for those pairs that are rotations of each other, the response time is proportional to the angle of rotation—that is, the angle of rotation needed *in 3D*.[6] That is a very remarkable ability, and a mind that can do it must be acknowledged to have some impressive ability to do actual mental rotations. That leaves unexplained what the "space" of the rotations is and whether, with appropriate brain-imaging techniques, one might develop mental rotations as a physical movie. But the inference to a mental-rotation facility of some sort in the mind seems inescapable. Perhaps our confidence in the conclusion is supported by our introspective awareness of mental imagery (some of the uses of which are discussed in the next chapter), but Shepard and Metzler's work itself needs no support from introspection—the evidence consists just of response times and their correlation with angles of rotation. (That is not to suggest that introspection and consciousness are beyond the pale of science; we will consider that issue in chapter 14.)

Once the method of inference from behavior to internal mental structure was in place, computer models of the mind proved valuable for keeping the inferences honest. Cognitive psychology came to be complemented by a wider field called "cognitive science"—a development not greeted with enthusiasm by some psychologists who called cognitive science "N disciplines in search of a grant." What cognitive science essentially added to strict psychology was computer modeling, on the grounds that if a mental phenomenon could be imitated by computer software, there was good reason to believe it had been understood (and if not, there was still a way to go).[7] Behavioral capabilities suggest an internal mechanism, then a hypothesized internal mechanism can be cloned into a software implementation, and then the output of that software can be compared to human performance. There have been some successes, sometimes in unexpected places—for example, the data-rich classification abilities needed to recognize faces can be imitated by "neural net" or "machine learning" software that is trained on data rather than being directly programmed.[8] In other areas, including ones we most care about, the insights from computer models have revealed instead how hard the problems are and how incomprehensibly brilliant the brain must be. Sight, for example, is something that psychology and neuroscience can discover a great deal about, but until it is possible for a computer to take in a raw image of a scene and print out a list of the objects in the scene—and that is still a long way off for natural scenes—then we cannot pretend we really understand how vision works.[9] Language translation is in much the same state. Anyone who has tried to use software to translate a Web page automatically soon finds that it mostly leaves clues requiring the reader to do the understanding, rather than being able to translate intelligently itself.

Baby Science: The "Scientist in the Crib"

Among the most fascinating bodies of knowledge about knowledge that psychologists have come up with is the science of baby learning.

Babies are very smart. They are born knowing very little, and within a couple of years they are walking and talking, activities that require intelligence barely matched by AI's 50 years of collective effort, vast grants, and massive computing power. They need to solve all the classic philosophical problems—the Existence of the External World, the Problem of Other Minds, Personal Identity, Induction and Evidence, How Does Language Mean? Why Should I Behave? and so on.[10] They need to solve those simultaneously with answering a host of scientific questions, like where food comes from and what characteristics of objects are dangerous. At the same time, they must use the results to take action, such as standing up and asking questions. If scientists knew babies' learning algorithms and could imitate them, they would be able to make science advance automatically and quickly.

Research has been intense in the last 20 years (that is, the researchers studied the babies intensively, though the reverse was also true). The most dedicated researchers rushed from lab meetings to maternity wards to be ready to videotape babies as soon after they were born as possible. Although it is not possible to ask newborn babies what they think, it is possible to make inferences by seeing what surprises them—babies exhibit a characteristic boredom response when their experience seems to be repeating itself and show a renewed interest with gaze fixation when they notice something surprising. Thus gaze fixation, carefully timed in videos, reliably indicates when babies notice that something has changed.[11] It therefore can be established that newborn babies have a basic counting ability, in the sense that they can tell the difference between two-syllabled and three-syllabled nonsense words.

CAN NEWBORN BABIES COUNT?

"As early as four days of age, a baby can decompose speech sounds into smaller units—syllables—that it can then enumerate.... Ranka Bijeljac-Babic and her colleagues at the Laboratory for Cognitive Science and Psycholinguistics in Paris have babies suck on a nipple connected to a pressure transducer and a computer. Whenever the baby sucks, the computer notices it and immediately delivers a nonsense work such as "bakifoo" or "pilofa" through a loudspeaker. All the words share the same number of syllables—three, for instance. When a baby is first placed in this peculiar situation where sucking yields sound, it shows an increased interest which is translated into an elevated sucking rate. After a few minutes, however, sucking drops [as the baby becomes bored]. As soon as the computer detects this drop, it switches to delivering words with only two syllables. The baby's reaction: It immediately goes back to sucking vigorously in order to listen to the new word structure. To ensure that this reaction is related to the number of syllables rather than to the mere presence of novel words, with some babies novel words are introduced while the number of syllables is left unchanged. In this control group, no reaction is perceptible. Since the duration of words and the rate of speech are highly variable, the number of syllables is indeed the only parameter that can enable babies to differentiate the first list of words from the second."

Stanislas Dehaene, *The Number Sense: How the Mind Creates Mathematics* (London: Allen Lane, The Penguin Press, 1997), 51, summarizing R. Bijeljac-Babic, J. Bertoncini and J. Mehler, "How Do 4-day-old Infants Categorize Multisyllabic Utterances?" *Developmental Psychology* 29 (1993): 711–721.

Baby interests are different—for example, they have an intense passion for looking at stripes—and to help them make sense of the sensory input, they come equipped with certain innate ... if not innate knowledge, exactly, then at least innate expectations that quickly lead to knowledge, such as a willingness to infer a world containing three-dimensional rigid objects and an ability to respond to people differently from objects that are not people. It has become clear that covariation plays a crucial role in the powerful learning algorithms that allow a baby to make sense of its world at the most basic level—in identifying continuing objects for example. Infants pay attention especially to "intermodal" information—structural similarities between the inputs to different senses, such as the covariation between a ball seen bouncing and a "boing boing boing" sound. That covariation encourages the infant to attribute a reality to the ball and event (whereas infants tend to ignore changes of color and shape in objects).[12]

Soon, infants become intuitive scientists, learning new concepts to make sense of experience, forming hypotheses and observing the predictions they imply, making interventions to see what the result will be.[13] And, eventually, some ex-babies whose curiosity has not shut down under the rush of adolescent hormones and parental pressure to acquire a career in law or accountancy actually turn into scientists.

CHAPTER 7

Mathematics

Mathematics is the gold standard of knowledge. Mathematics is proved true, and it stays proved true. Proofs convey certainty and (if the proofs are not too long for our minds to cope with) they induce understanding of *why* mathematical truths *must* be true.

Let us take a very quick example, leaving a few more elaborate ones for later.

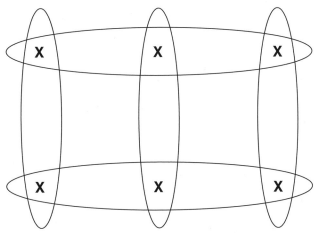

FIGURE 7.1   Why $2 \times 3 = 3 \times 2$.

The two rows of three objects must have the same total as the three columns of two, because they are just the same objects, but divided into parts differently. The proof applies equally to the actual six marks on the page as to any other array of six objects, physical or abstract.

Of course, not everyone is happy. If there was ever an age to take a gift horse lying down, ours is not it. Surely certain knowledge about reality with complete understanding is a promise too good to be true? Attacks have come from all sides. An old theory of mathematics, Platonism, held that knowledge of mundane reality can only be uncertain, so that if mathematics is certain it must be about some other realm beyond space and time, inhabited by abstract objects like Numbers and Sets. That left it very mysterious how mathematical truths could be so useful in the real world we live in. Another view is expressed in Bertrand Russell's statement: "Mathematics may be defined as the subject where we never know what we are talking about, nor whether what we are saying is true."[1] On that view, mathematics does not have a subject matter at all but is only about deducing one thing from another. Some take an "if-then-ist" view, asserting that one may choose any axioms one likes and that mathematics is merely about showing what follows from which axioms, that is, that mathematics is a part of logic. Many scientists and engineers who use mathematics as a tool have taken a similar view of it as merely a calculating device or language or "theoretical juice extractor," a grab bag of methods for moving from one empirical scientific fact to another. Einstein said, "As far as the propositions of mathematics refer to reality, they are not certain; and as far as they are certain, they do not refer to reality,"[2] asserting a gap between the mathematical and physical realms reminiscent of Platonism. In more recent times, there have been attempts to cast doubt on the certainty of proof, beginning with Imre Lakatos' *Proofs and Refutations* and concluding with the usual sociological and postmodernist attacks that have plagued science in general.

None of those views of mathematics are correct. That two rows of three dots have the same number of dots as three columns of two is a truth about real dots on paper. It is not about deriving one truth from another, it is about dots. It cannot be made false by choosing different axioms. One can choose any axioms one likes and call the objects satisfying them "numbers" if one chooses, but that will not affect real arrangements of dots, which must conform to the truth $2 \times 3 = 3 \times 2$. And whatever may or may not be true in Platonic realms of numbers, arrangements of dots are in this world and the facts about them are perceived to be true in this world.

Contrary to Russell's view, mathematics does have a subject matter, an aspect of the real world that it studies, in exactly the same sense as biology studies the living aspects of the world. The subject matter of mathematics is structure, or pattern.

Symmetry

The words "structure" and "pattern" are unfortunately too vague, general, and colorless to bring immediately to mind any clear picture of what mathematics is about. We will look at two of the most easily appreciated yet most abstract objects of mathematics, *symmetry* and *ratio*.

Dinner tables and the human body have symmetry (not always exact), while palindromes have a perfect symmetry. Symmetry is a property at once highly abstract—much more so than shape and size, for instance—yet at the same time an easily perceived property of real things. If you are considering a career in politics and have a slight asymmetry in your face, either invest in plastic surgery or change your plans, because it is immediately obvious on television and humans are wired to react badly to it.

There is not a great deal to say mathematically about simple bilateral symmetry. The action begins when an object has several symmetries. A square, for example, is symmetrical about a vertical axis, a horizontal axis, and both its diagonals. A body is said to be symmetrical about a point p when a point is in the body if and only if the point directly opposite it across p is also in the body. Thus a square is symmetrical about its center. The following is a necessarily true statement about real bodies: All bodies symmetrical about both a horizontal and a vertical axis are also symmetrical about the point of intersection of the axes.

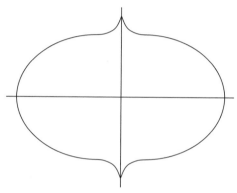

FIGURE 7.2   A figure vertically, horizontally, and centrally symmetric.

That is because the central symmetry is the *composition* of the vertical and horizontal symmetries. We find ourselves forced to use the word *symmetry* not just of the property that the body has, but of the transformation or operation of reflecting or rotating the object in a way that brings it back into coincidence with itself. Thus the composition of a vertical and a horizontal reflection is "turning the object inside out" about the point of intersection of the axes (or, what comes to the same thing, rotating it 180° about the point).

The study of the composition of such operations is the beginning of group theory, the central topic in abstract algebra.[3] It begins to show its power when we consider one of the most symmetrical of easily visualizable objects, the cube.

Exercise: how many rotations of the cube are there (that is, rotations that move the cube so that it occupies the same space)?

There are the rotations of 90°, 180°, and 270° around the axis through the center of each face, and ... (answer at the end of the chapter).

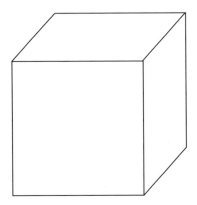

FIGURE 7.3   How many rotations of the cube are there?

Measurement, Quantity, and Ratio

All perfect circles are alike; each imperfect circle is imperfect in its own way.

One way in which all perfect circles are alike in that they have exactly the same ratio of circumference to diameter: for any perfect circle, it is exactly π times as far around as it is across. The number π (a little over 3, about 3.14159) is a useful test bed for explaining many mathematical concepts—even induction, as we saw in chapter 1 and will see more fully in the next chapter. But let us first concentrate on its character as a ratio or proportion. Ratio, like symmetry, has the character of being highly abstract but at the same time realizable directly in physical things. That gives it a crucial role in measurement.

Twentieth-century discussions about mathematics, mathematical knowledge, and the relation of mathematics to the world were largely silent about measurement. That is a very strange absence, since the naïve user of numbers in, say, accounting or woodwork is under the distinct impression that measurement is the chief way to find mathematical structure in the world. You lay a ruler along a length of stone and read off a number, which tells you how that stone compares in length with other stones, and leads to predictions about whether it will fit in a wall. That was, indeed, the point of view of the longest-running philosophy of mathematics, the one that dominated the subject from ancient times to the nineteenth century. According to the Aristotelian view, mathematics is the "science of quantity." Quantity is a property that physical things have, and the way to find out about quantity is to count (if the quantity is discrete) or measure (if it is continuous). In the most straightforward continuous cases, like length, one chooses a unit arbitrarily and measures the ratio of all other lengths to the unit. ("By *Number*," Newton says, "we understand not so much a Multitude of Unities, as the abstracted Ratio of any Quantity, to another Quantity of the same kind, which we take for Unity."[4])

Although there is choice in the unit, there is no choice in the system of ratios: the ratio of any length to any other length is an objective feature of the world. *We* have only a choice of what names to call lengths—but only one choice: once we have determined which length to call the unit, our numbering system gives names to all other determinate lengths.

It is true that there are some other quantities where numbers can be assigned but the numbers do not express ratios. With hardness of minerals, IQ, temperature in Fahrenheit, and others, what is objective is just the *ordering* of the quantities. That is why it does not make sense to say that 20°F is twice as hot as 10°F, although it is objectively true that 20°F is somewhat hotter than 10°F. (The absolute temperature scale, starting at −460°F, does provide a kind of ratio scale of temperature, but one based on elaborate theory and distant from ordinary experience.[5])

THE SCIENTIFIC REVOLUTION'S LAWS OF PROPORTION

Kepler's Second Law: The area swept out by a radius from the sun to a planet is proportional to the time taken (1609)

Snell's Law: When light is refracted at a surface, the sine of the angle of refraction is proportional to the sine of the angle of incidence (1602, 1621, 1637)

Galileo's Law of Uniform Acceleration: The speed of a heavy body falling from rest is proportional to the time from dropping (1638)

Pascal's Law: The pressure in an incompressible fluid is proportional to depth (1647)

Hooke's Law: The extension of a spring is proportional to the force exerted to stretch it (1660)

Boyle's Law: For a fixed quantity of gas at constant temperature, pressure is inversely proportional to volume (1662)

Newton's Second Law of Motion: The acceleration of a body is proportional to the total force acting on it (1687)

Newton's Law of Gravity: The force of gravity exerted by one body on another is proportional to the masses of each and inversely proportional to the square of the distance between them (1687)

Newton's Law of Cooling: The rate of temperature loss from a body is proportional to the difference in temperature between the body and its surroundings (1701)

Ratio and symmetry are just two examples of the structures or patterns that mathematics studies. Before exploring some of the great themes of mathematical structure—discrete and continuous, local and global, simple and complex—let us pause to consider the special contribution that mathematics has made to methods of *knowledge*, proof.

Proof and Understanding

Proof is what makes mathematics different from other sciences.

In physics or biology or economics, results come only after hard work with observations and experiments. The raw experience may need theory and mathematics to cook it, but the ingredients are harvested from the wild. In mathematics, there is proof by pure thought instead. Let us start with an example more typical than the "at a glance" example above of why $2 \times 3 = 3 \times 2$.

How can you add up the numbers 1 to 100 (in some sensible way that gives understanding of the result)? Write down:

$$1 + 2 + 3 + 4 + \ldots + 98 + 99 + 100$$

Now write underneath them the same numbers, backward:

$$1 + 2 + 3 + 4 + \ldots + 98 + 99 + 100$$
$$100 + 99 + 98 + 97 + \ldots + 3 + 2 + 1$$

Next add each number in the top row to the one directly below it. It is clear that each pair adds up to 101. There are 100 pairs. So the sum of all the numbers written down is 100×101. This is twice the sum of the numbers 1 to 100 (since we wrote that sum down twice). So

$$1 + 2 + 3 + 4 + \ldots + 98 + 99 + 100 = \tfrac{1}{2} \times 100 \times 101$$

which is 5,050. The answer has been *proved* correct.

A proof has three advantages over simply finding the answer by putting the numbers into a calculator or slaving over a large piece of paper:

1. You *understand* the whole process completely, and hence achieve that sense of intellectual power that is one of the payoffs of studying mathematics.
2. As a result of your understanding, you have *certainty* that the result is correct. There is no prospect of a mistake in

entering the numbers, or of the kind of measurement and observational errors endemic in the other sciences.

3. Also because you understand, you can *generalize*: it is clear that the fact that the last number happened to be 100 was incidental to the proof. It is just as easy to prove, with the same method, that for *any* whole number *n*,

$$1 + 2 + 3 + 4 + \ldots + n = \tfrac{1}{2} \times n \times (n+1)$$

If anything, the problem is easier for arbitrary *n* than for 100, since there is no chance of being distracted by any particular facts about the number 100. So proving an infinite number of facts can be easier than proving one.[6]

Understanding can be enhanced by diagrams, even though they are traditionally not counted as real proofs in the serious mathematical literature (for good reason, in that they can be misleading in ways that symbols are not). It is natural to picture the problem of adding the numbers 1 to *n* thus.

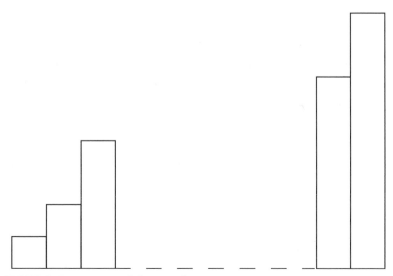

FIGURE 7.4   Sum of the numbers 1 to *n*.

Now imagine taking a copy of the diagram out of the page, turning it over, and placing it above the original diagram, as indicated below with the dotted boundary.

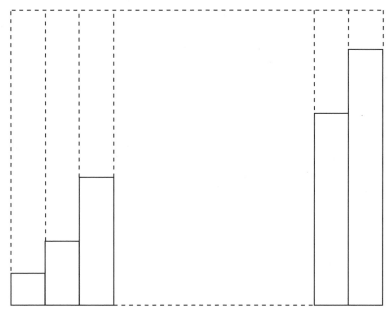

FIGURE 7.5 Twice the sum of the numbers 1 to n.

It is clear that the large rectangle formed has width n and height $n + 1$, and so the area of the rectangle formed is $n \times (n + 1)$. The sum of the numbers 1 to n is half this, yielding the same answer as before.

It is also clear that this proof is just a geometrical version of the symbolic proof above. It might be easier to grasp the picture intuitively, but it would be simple to translate it into words and symbols, if desired.[7]

Of course, proof is not always as easy as that ... Sir Walter Raleigh asked his mathematical adviser for a formula for the number of cannonballs in various stacks on the deck of his ship. The adviser came up with something, but realized it was not clear whether the balls were being stacked in the most efficient way. There is a natural way to pack spheres tightly: lay down the deck layer in "hexagonal" fashion, where each ball is surrounded by six others. Then put the second layer so that the balls lie in every second depression in the first layer; the second layer will then also form a hexagonal packing. Then put the third layer in every second depression of the second layer, so that the balls of the third layer are directly above those of the first. If this is done on a large scale, the balls will occupy $\pi/\sqrt{18}$ of the space, or just over 74 percent.

Was this the densest packing possible? Kepler conjectured that it was. Fruit-shop owners thought it was obvious. A New Zealand greengrocer was amazed to hear that mathematicians had been working on the problem for 400 years.[8] Asked how long it took him to find the best packing, he said "You just put one on top of the other. It took me about two seconds." But conjecture and tradition are not enough for mathematicians. They demand proof. A proof now exists, but it was not achieved until 1998, when Thomas Hales of the University of Michigan put together the last steps, using computer verifications of many complicated cases.

That is, Hales credibly claimed to have done so. In 2003, the head of the refereeing panel announced they were "99 percent certain" of the correctness of the proof, but they could not certify the correctness of all of the computer calculations. Hales announced a collaborative project to check all the computer calculations with automatic proof-checking software, but it may take 20 years.[9]

Spheres are very simple shapes, of course. Shortly after Hales announced his proof of the Kepler conjecture, he had a call from the Ann Arbor farmers' market. "We need you down here right away," they said. "We can stack the oranges, but we're having trouble with the artichokes."

Discrete and Continuous

The ancient view of mathematics, that it was the "science of quantity," went on to classify quantity into two kinds: the discrete, studied by arithmetic, and the continuous, studied by geometry. The contrast between the discreteness/atomicity of the whole numbers and the continuity of such quantities as space, time, and motion is one of the great themes of mathematics.

The Greeks discovered two fundamental truths about the continuous and the discrete, which have been central to the direction in which mathematical knowledge has developed ever since. The first was their fundamental distinctness. Perhaps the first truly surprising result in mathematics was the one attributed (traditionally but without much evidence) to Pythagoras, the proof of the incommensurability of the side and diagonal of a square. It is natural to think that it is possible to convert any continuous quantity to a discrete one by choosing units on a ruler. Given a ruler divided finely enough, it should be possible to compare any continuous quantities, say, lengths, by counting exactly how many times the ruler's unit is needed to measure each quantity. One length might be 127 times the unit and another 41 times, showing that the ratio of the lengths is 127 to 41. Given any two lengths (not yet divided into units), it is possible to find the largest unit that "measures" them (that is, of which they are both whole number multiples) by a process the Greeks called *anthyphairesis*.

FIGURE 7.6   First step in the *anthyphairesis* of lengths A and B.

Given two lengths A and B, we see how many times the smaller one (say B) fits into the larger one A (in the example, 3 times). If it does not fit exactly a whole number of times, there is a remainder R that is smaller than both A and B. (If B does fit exactly, then of course B itself is the unit that measures both A and B.) Any unit that measures both A and B must also measure R (since R is just A minus a whole number of Bs). So we can repeat the process with R and B, either finding that R measures B (and hence A as well), or that there is a smaller remainder R′, which must also be measured by any unit that measures A and B. And so on. Since we always get smaller remainders at each step, we work our way down until the last remainder is the unit that measures all previous remainders and hence also measures A and B.[10]

Now, what happens if we apply *anthyphairesis* to those two very naturally occurring lengths, the side and diagonal of a square? The side fits once into the diagonal, with a remainder left over, which we can lay off against the side, and…

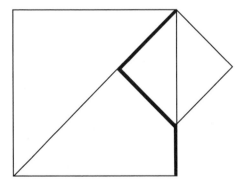

FIGURE 7.7 *Anthyphairesis of diagonal and side of a square.*

The first remainder (diagonal minus side) is the length drawn in thick lines. It appears several times in the diagram. It fits twice into the (original) side, and when we take the (small) side length out of the (small) diagonal, we are in the same position as we were originally with the larger square: taking a side out of a diagonal. Thus the small square, with its diagonal, is a repeat of (the same shape as) the large square with its diagonal, so *anthyphairesis* goes into a loop and keeps repeating: at each stage, one side-length is taken out of one diagonal. Therefore, the remainders just keep getting smaller and smaller, and the process never ends. There is thus *no* unit that measures the original diagonal and side. The diagonal and side of a square are "incommensurable." So the ratios of continuous quantities are more varied than the relations of discrete quantities. Therefore geometry, and continuous quantity in general, are in some fundamental sense richer than arithmetic and not reducible to it via choice of units. While much about the continuous can be captured through discrete approximations, it always has secrets in reserve.

The second major discovery of the Greeks was that the continuous sometimes naturally gives rise to discrete structures. Their classification of mathematics included not only pure geometry and pure arithmetic, but three "subalternate" or applied mathematical sciences, astronomy and optics (subalternate to geometry), and music (subalternate to arithmetic). The reason music was considered allied to arithmetic was the discovery, also attributed to Pythagoras, of the connection between the harmony of notes and the whole-number ratios of the lengths of the strings producing them. Although the lengths of strings and the pitch of notes are both quantities whose nature is to vary continuously, a discrete pattern emerges from them: for strings of a fixed tension, the easily perceived harmonies of octave, fifth, and so on are produced by pairs of strings whose lengths are in small whole-number ratios. The simplest non-trivial ratio, 2:1, produces the most prominent harmony, the octave. Later science has explained the mystery and at the same time found one mechanism for how continuous variation can produce discreteness: pitch is caused by the frequency of soundwaves, and there is something special about the frequencies of waves in small integer ratios. They interact to produce regular large peaks that affect the eardrums; randomly attuned frequencies do not.

The interplay between the discrete and the continuous has been at the heart of some of the most profound advances in science. We saw briefly in chapter 1 how the discovery of small integer ratios in the amounts of interacting chemicals provided the original stimulus for Dalton's atomic theory of matter. At a deeper level, the discreteness of the naturally occurring elements (and their isotopes) is now understood to be a consequence of the discrete array of solutions of the Schrödinger equation.

A more easily appreciated example is Mendel's work on genetics. One of the many problems with Darwin's original theory of evolution concerned "blending inheritance." Natural selection, according to Darwin, acts on small mutations that occur rarely and at random in populations. But since the possessor of such a mutation has to breed with the "normals" of the population, surely a favorable mutation should blend back into the population over a few generations before natural selection has a chance to act? Like many other problems with evolutionary theory, it was only admitted to be serious once it had been solved. The solution came with Gregor Mendel's discovery that the genetic material was, often at least, discrete. And the discovery of discreteness, through Mendel's experiments with peas in his monastery garden, came, as it had with the Pythagoreans and Dalton, with the noticing of small whole-number ratios in material that was in its nature continuous.

Mendel took two lines of peas, short and tall, each bred over several generations to be "pure" or "true" (that is, the short ones produced only short ones and the tall ones only tall ones). Then he crossed them, that is, pollinated the flowers of the short ones with pollen from the tall ones. With a genius for "population thinking" exceedingly rare in his age, he thought to perform this experiment many times and count the results. He found, not very excitingly, that all of the crosses were tall (not somewhere between short and tall, but clearly tall). But if this generation of tall ones were crossed with one another, the results were not all tall but consistently about three-quarters tall and one-quarter short. The simple phrase "consistently about three-quarters ..." is deceptive. It makes something hard to find look easy. The discreteness of the ratio 3:1 does not reveal itself in any single experiment, especially a small one with one or two plants. One must take a count in a large number of experiments, and be content that the ratio is not exactly 3:1.

Mendel correctly posited a theory of discrete factors or genes that would explain the simple ratio. He supposed that length of peas depends on each individual's having two genes for height, each either "tall" or "short," with the combination "tall" and "short" resulting in the same appearance as two "talls," and with random assortment of genes between generations. When first-generation plants (all appearing tall but in fact all having one "tall" and one "short" gene) are crossed with one another, the next generation has about one-quarter "tall-tall," one-half "tall-short," and one-quarter "short-short"; only the last one-quarter appear short.

With one bound, Darwin's theory was free of the blending-inheritance problem. An assortment of discrete genetic material is capable of a giving a mutation several chances to express itself fully. How well Darwinism has managed to escape several other problems remains to be seen (chapter 13).

There have been widespread doubts as to whether Mendel's published results were *too* close to the 3:1 ratio to be convincing. His personal papers do not survive so it is hard to know for sure. The results he claimed for the tall-short experiment were 1,787 tall to 227 short, or 2.84:1. That is not very close to 3:1, and indeed would only be recognized as a 3:1 ratio in the context of several other similar results. Defenses of Mendel have appeared, but to little effect.[11] "Hero's feet found not of clay after all"—what kind of a headline is that?

Mathematical Models: Local and Global Structure

Other than measurement, the main way in which mathematics applies to the world is through the process of mathematical modeling.

Recall how compound interest works. If money is invested in a bank at 2 percent per month compound interest, the accumulated amount (principal plus interest) after t months, P_t, is related to the amount of the month before, P_{t-1}, by

$$P_t = P_{t-1} + {}^2/_{100}\, P_{t-1}$$

The formula says, "each month, add to the accumulated amount 2 percent of itself to get next month's amount."

That equation expresses the *local* structure, the connection between the amounts in consecutive months. The bank's computer starts out with the original principal, and goes through step by step, using the equation to calculate the accumulated amount after t months. The resulting *global* structure, the general shape of what happens over time, is represented by the familiar rising exponential growth curve.

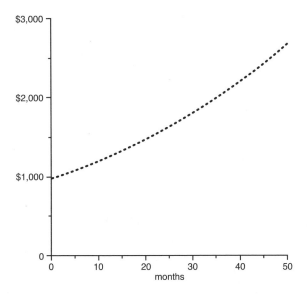

FIGURE 7.8 Growth of $1000 at 2% per month compound interest.

Similar phenomena arise in continuous cases such as the exponential growth of populations. These are *models* in a stricter sense, in that the aim is to find the mathematical structure of a preexisting natural phenomenon, whereas in the compound interest case the bank imposes the formula by fiat. Populations are in actuality discrete (with individuals born and dying at random times), but for large populations it is simpler to iron out the small blips and approximate by regarding the population as well as time as continuous. If a population grows continuously at, say, 2 percent a month, then again some mathematical reasoning can extract the global structure from the local, and show that, over time, the result is the familiar exponential growth curve. Of course, its shape is similar to the compound-interest graph except for being continuous.

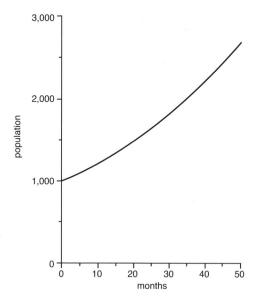

FIGURE 7.9 Continuous growth of a population at 2% per month.

Again, the exponential shape is not a property of the originally given local structure, the annual growth rate. It is visible only in the global structure.

Further mathematical reasoning can uncover such global features as the rule of thumb: if a population grows at x percent per year, the number of years for the population to double is found my dividing x into 70 (if x is fairly small): thus 2 percent growth, 35 years to double; 10 percent, 7 years to double.

These features are common in mathematical models generally.[12] The computer simulation of, say, the growth of a city, will exhibit phenomena explainable as the results of gradual accumulation of interactions among its parts, the details depending on the assumptions made about, for instance, the impact of siting factories near a residential area on the medium-term development of the area.[13]

A model of population growth or urban interaction is all very well, but does it *model* the real situation? Do populations or cities really grow like that?

Sometimes they do, sometimes they don't, and there is no substitute for measuring real growth rates. Studying real phenomena by mathematical modeling involves measurement and observation, as well as purely formal work. Paul Ehrlich's 1968 bestseller, *The Population Bomb*, confidently extrapolated world population growth to the near future and concluded "in the 1970s and 1980s hundreds of millions of people will starve to death." But it was already clear, or should have been, that birth rates had begun to drop precipitously in most countries. More complicated population models can account for varying growth rates, and the results need not be anything like exponential growth. Current UN projections estimate that zero population growth for the world will be reached about 2050, followed by a decline.[14]

Such mistakes, and the term "model" itself, draw attention to the model-reality gap of which Einstein spoke. Talk of modeling can suggest that the practice involves studying not the real world but something distinct from it, an abstract structure called a "model." That suggestion is misleading because, although there is a gap, the model has properties that the real world can share. Just as a model ship can literally have the same shape (approximately) as a real ship, a mathematical model can incorporate relevant real world structure.

This is most clearly evident in the rare but important cases where the model is not just some kind of abstraction or idealization or approximation of the real thing, but an exact fit. The model really does contain all the relevant structure of the reality it models. A classic case is Euler's problem of the bridges of Königsberg.

In the eighteenth century, the city of Königsberg had a river with banks and two islands connected by seven bridges, as shown.

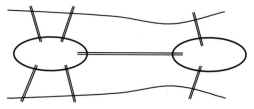

FIGURE 7.10   The Bridges of Königsberg.

The citizens noticed that it was impossible to walk over all the bridges once, without walking over at least one of them twice. The most celebrated mathematician of the century, Leonhard Euler, *proved* mathematically that they were right: it is absolutely impossible to walk over all the bridges once and once only. (His proof relies on the fact that if one walks in and out of a region—bank or island—one "uses up" two bridges, and since the number of bridges leaving each region is odd there will be some left over; but a conceptually simpler proof could just laboriously list all the paths that do not repeat any bridges, and check that none of them crosses all the bridges.)

In this case, the "model" is the diagram above, which simply states what regions are connected by what bridges. That is *all* the information about the real situation that is relevant to the problem—the shapes of the islands, lengths of the bridges, and so on have no relevance. Thus the model is a complete picture of the structure of reality that is relevant to the problem. There is no "model-reality gap" that requires us to use judgments about degrees of approximation or idealization. Euler's proof allows us to know something with certainty about the structure of the real system of islands and bridges, not just about some abstraction of it.

Euler's example is important to mathematics for another reason. It shows that the old view of mathematics as the "science of quantity" is too narrow. Euler begins his article with the comment, "In addition to that branch of geometry which is concerned with magnitudes, and which has always received the greatest attention, there is another branch, previously almost unknown, which Leibniz first mentioned, called the geometry of position. This branch is concerned only with the determination of position and its properties; it does not involve measurements…."[15] Although elementary mathematics is mostly about quantities (whole numbers, ratios), Euler's result is about pure arrangements. It is now regarded as the first result in network topology, one of the forerunners of the explosion of "formal" or mathematical sciences that are the subject of chapter 9.

PDEs: Simple and Complex

Those who have a basic acquaintance with science are often familiar with (ordinary) differential equations, which describe how some quantity develops with time—a large population grows at 2 percent a year, a planet moves in an orbit determined by the sun's gravity, a falling heavy body accelerates uniformly. Much less familiar, but equally important and with a distinctly different mathematical character, are partial differential equations (PDEs). These typically describe systems whose development in time is driven by spatial variation. Wind, for example, is driven by flow of air from areas of high pressure to low pressure: the spatial variation of air pressure is what drives the pressure itself to vary as air flows. The wind direction (at each point) is determined by the direction in which pressure varies from high to low, and its speed is determined by how quickly air pressure changes in that direction. The faster the air pressure varies over space (the more closely bunched the isobars), the more force-fully wind is sucked from high to low.

It is similar with heat in a solid body, which diffuses from hot to cold, and diffuses more quickly the faster the spatial transi-tion is from hot to cold. The "heat equation" is one of the classic PDEs. Another is the "wave equation," which describes how a disturbed surface "pulls back" toward equilibrium, pulling back more strongly the further it is away from being flat; the solu-tions of the wave equation are the actual waves generated by this mechanism. As we saw earlier, PDEs are also fundamental to the physics of electromagnetic radiation, relativity and quantum mechanics.

PHYSICS STUDENTS' T-SHIRT: MAXWELL'S EQUATIONS OF ELECTROMAGNETISM

And God said...

$$\nabla \cdot E = \rho/\varepsilon_0$$
$$\nabla \cdot B = 0$$
$$\nabla \times E = -\partial B/\partial t$$
$$\nabla \times B = \mu_0 J + \mu_0\varepsilon_0 \partial E/\partial t$$

... and there was light.

There are good PDEs and there are bad PDEs. The difference provides a nice illustration of another of the great themes of mathematics, simple versus complex. The heat equation is good, the Navier-Stokes equations that govern large-scale weather are bad.

What is good about the heat equation is its smooth and predictable development. It is easy to imagine how diffusion works. To illustrate, we choose an example restricted to one spatial direction, a rod infinite in one direction, initially cold, then with its one end suddenly heated to a high temperature, and, finally, with the heat removed. An intuitive understanding of what profile of temperature should be expected at later times is sufficient for a basic insight into the PDE for heat flow. Initially, all the heat is at zero; later, it gradually spreads along the rod, as illustrated by these graphs of the temperature profile in the rod at three later times:

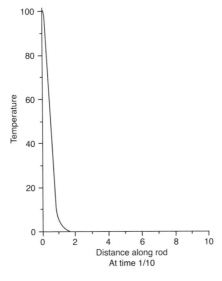

Distance along rod
At time 1/10

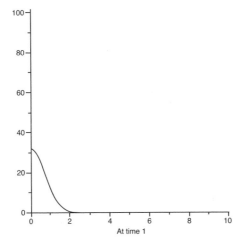

At time 1

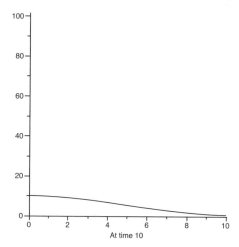

At time 10

FIGURE 7.11 Successive temperature profiles when heat diffuses along a rod.

(The shapes of the graphs are the bell curves familiar in statistics.)

All is calm, smooth, and as it should be. The averaging process gradually "blurs" any initial profile of temperature, by repeatedly replacing the temperature at a point by an average of the temperatures at nearby points.[16]

This tractability has proved important in the application in recent decades of mathematical methods to finance. The Black-Scholes PDE governs the price of options (for example, if you offer me the right to buy 100 shares in Texas Instruments at $30 in a year's time, given we both know the current price, $28, and the "volatility" or jumpiness of the stock, what is a fair price?). The equation was solved around 1970, in work that won the authors the 1997 Nobel Prize in economics. The solution is the Black-Scholes option pricing formula that is said to be the world's most-used mathematical formula (because owners of options have the formula on their computers continually repricing them as the price of the stock changes). After the solution was found, it was realized that it should have been easy, since the Black-Scholes PDE is just a minor rescaling of the heat equation—but with the time backward.

That is how a good PDE behaves. On the dark side, the PDEs governing fluid flow behave very badly. If the heat equation is among the quiet rabbits of PDE fauna, the Navier-Stokes and similar equations that govern fluid flow are the rogue elephants. That is problematic for any attempts to model fluid flow: the weather, the flow around a submarine hull, or the global circulations that determine world climate changes. Though there is such a thing as quiet and smooth fluid flow, it is not typical. Fundamental to most fluid flow is the phenomenon of turbulence. As soon as a smooth flow speeds up, eddies and turbulence appear. Points in the flow that were close are quickly forced apart into very different, complex trajectories. Although turbulent flow is just as strictly deterministic and rule-governed as smooth flow or heat diffusion, prediction of an individual trajectory is practically impossible beyond small time intervals. Although at the local level those PDEs look much the same as the heat equation, in that they tell each point "where to go next," the global effect as the small changes accumulate is very different.

Those problems proved insurmountable for the first serious attempt to predict the weather by calculation, Lewis Fry Richardson's *Weather Prediction by Numerical Process* of 1922. After explaining correctly the equations governing the weather and how a discrete approximation to them could be made to actually achieve predictions, Richardson illustrated his work by predicting, or rather retrodicting, the weather at 1 p.m. on May 20, 1910, from the state of the atmosphere at 7 a.m. that day.

The day was a data-rich one, as many balloons had been sent aloft to observe whether there were atmospheric effects from the earth's passing through the tail of Halley's comet. (There weren't.) Richardson divided Western Europe into a grid of squares about 200km × 200km, with the atmosphere divided into five vertical levels. In each "cube," the 7 a.m. pressure and wind was set, as observed. The Navier-Stokes equations (as applying to fluid flow with friction and gravitational effects and allowing for the earth's rotation) then allowed the (approximate) calculation of the pressure and wind at later times—Richardson took time steps of three-quarters of an hour. That is a great deal of calculation by hand, which Richardson accomplished over a long period in the intervals of his work with an ambulance unit on the Western Front. He arrived at a prediction of the pressure in the central area of the grid, around Munich, for six hours after the initial time. The result could then be compared with the known reality.

His answer was absolutely and totally wrong. The calculation predicted a huge, indeed physically impossible, rise in pressure whereas the actual weather was rather stable and there was hardly any pressure rise. Unlike some scientists who publish only unfalsified predictions, Richardson made no attempt to fudge the result and freely allowed that it was wrong. There was nothing seriously mistaken in his methods, calculations, or data. The culprit was the behavior of the equations. Air pressure is subject to oscillations of periods around one hour, so the actual pressure readings at 7 a.m. were rather variable around their mean or "true" values. The equations amplified those divergences from true, resulting in nonsense only a few time steps ahead.[17]

Modern computers and methods for smoothing data and ironing out some of the theoretical problems have brought weather prediction by computer models within the bounds of possibility. As can be seen by saving the TV weather reports and checking them a few days later, the results are generally usable but far from perfect.

In 2000, the Clay Mathematics Institute named prizes of $1 million each for the solving of seven outstanding mathematical problems. The money will be paid for a solution of one of such conjectures as the Riemann hypothesis or the P=NP problem— or for "substantial progress" on understanding the solutions of the Navier-Stokes equations.

It is not recommended as one of the easier methods of becoming a millionaire.

(Answer to the earlier question: There are 23 rotations of a cube—24 if you count "rotation through 0°," or doing nothing. Besides the ones around the axes through the centers of the faces, there are ones around axes connecting the midpoints of opposite edges, and about axes connecting diametrically opposite vertices. Now find the "multiplication table" of these rotations: for example, what is the result of combining two rotations of 90° about axes through the centers of two different faces?)

CUBE VISUALIZATION EXERCISES

1. Imagine looking down on a (opaque) cube from above one of the corners. How many corners are visible (including the ones outlined against the background)?
2. Strike a soft surface, such as clay, with the corner of a hard cube. What is the shape of the (outline of) the impression in the soft surface?
3. Imagine a wooden cube painted on the whole of the outside. Cut the cube evenly into 3 in each dimension, forming 27 smaller cubes. How many of them are painted on exactly 2 faces?

CHAPTER 8

Enemies of Mathematics

It might be wondered what was happening in the mathematical arena while science faced its sustained attack from the irrationalists. The truths of mathematics have always been regarded as more certain and unchanging than those of science. So one would expect that the defenders of rationality should have been able to fall back on mathematics when rationality in general was under attack. One would also expect that the more shameless and foolhardy in the irrationalist camp should have been keen to make a name for themselves by trying to knock over this last line of defense.

Lakatos, Proofs, and "Proofs"

As in science, the attack was launched by someone who had a certain understanding of the subject and did not entirely intend for his work to be as eagerly taken up as it was in the cause of irrationalism. Imre Lakatos's work of the mid-1960s, published posthumously as *Proofs and Refutations*, used "rational reconstructions" of certain episodes in the history of mathematics in support of the thesis that mathematical proof is *fallible*.[1]

The idea of the book is to follow the historical development of the proof of a single mathematical theorem, Euler's theorem on polyhedra, and to "show" that it did not really succeed in establishing the theorem beyond doubt. One of several ironies in the affair was that the success of Lakatos's book was partly due to the fact that the theorem being undermined was so interesting. It is easy to appreciate even for those whose school mathematics has virtually disappeared, and is in any case quite unlike the recipes for symbol manipulation that make school arithmetic so boring in the first place.

Imagine a cube. (Or go back to the cube picture in the previous chapter, if you need to, but imagining it is better for you.) Count its faces (6: top, bottom and four sides), its edges (12: four around the top, four around the bottom and four down the sides), and its vertices or corners (8: four around the top and four around the bottom). Now calculate the number $V - E + F$, meaning the number of vertices minus the number of edges plus the number of faces. For the cube it is $8 - 12 + 6$, which is 2. The remarkable thing is that $V - E + F$ is also exactly 2 for almost any solid one can think of, even those which vast numbers of faces, in the style of a Buckminster Fuller dome. (Even more generally, the flatness of the faces has nothing to do with the result, so it works for a network of lines drawn on the surface of a sphere as well.)

A simpler case is a pyramid (including the bottom), for which $V - E + F$ is (please try out your mental visualization facility one more time) $5 - 8 + 5$, which is 2 again. The secret lies, somehow, in the minus in front of the number of edges E: for solids with many faces, the extra vertices and faces are balanced by the extra edges. To begin to understand why this balance is exact— why extra vertices and faces are always exactly balanced by extra edges—imagine (last time) a square face on some perhaps large solid. What happens if we draw a diagonal of the square? The number of vertices V does not change. F goes up by one (since the square face has become two triangles), and E goes up by one (since we have added one new edge, the diagonal). So V – E + F for the whole solid is unchanged. The standard proof of the theorem proceeds by using this idea in reverse: one gradually removes edges, faces, and vertices, showing that V – E + F is unchanged in the process, until one reaches a simple figure where one can count and find that V – E + F is 2.

So what does Lakatos have to say? He points out that there are certain solids for which the theorem is not true. One example is a cube with a hole drilled through it. *Proofs and Refutations* is structured as a classroom encounter between a teacher and a crowd of besieging students, who suggest further counterexamples to the theorem each time the teacher tries to amend the theorem to exclude them. There are many historical footnotes intended to show that the classroom encounter is a "rational reconstruction" of the real history of the theorem. Lakatos's conclusion is that this theorem, and, by implication, mathematical theorems in general, are never really *proved*, but are always subject to refutation by further counterexamples.

It needs mathematical advice to explain why Lakatos's project is fundamentally dishonest. Most of the many mathematicians who read *Proofs and Refutations*, it is true, were not very alarmed by it. But that is because of the touchingly mole-like blindness to matters philosophical for which they are known, and which led them to believe Lakatos's conclusion was merely that the teaching of mathematical theorems ought to be more interesting. Better informed mathematicians were not deceived. The distinguished Princeton mathematician John Conway points out that, for one thing, Euler's theorem is atypical in that the terms in it turn out to be difficult to define.[2] "Face" might seem clear, but do ring-shaped faces count? The vast majority of mathematical theorems—about numbers or algebra, for example—do not have any such problems. Lakatos also refused to admit that the theorem has been proved of *some* solids—for example cubes—until it is established for *all* the solids for which it is true. But most misleading of all was his failure to describe the final state of the theorem. All the problems about ring-shaped faces, holes, and so on can in fact be cleared up, but doing so requires that one should have the full classification of surfaces correct. This is a difficult and complicated, but not infinitely difficult and complicated, mathematical result. Lakatos stops his story just before getting to this result, in order to leave the impression that the truth will never be pinned down. It is simply impossible that he should have stopped at exactly that point honestly.

There is more to complain about. For example, there are doubts about one's claim to be supported by history, when one has admitted to writing a "rational reconstruction" of history, that is, the history that would have supported one if it had actually happened. And Lakatos is one of the worst offenders in the compulsive scattering of quotation mark to "neutralize success words." The word "proof" is a success word, that is, it is part of its meaning that something proved is true, and that the proof has succeeded in showing it is true. So when Lakatos has his teacher say things like, "I am not perturbed at finding a counter-example to a 'proved' conjecture; I am even willing to set out to 'prove' a false conjecture!" it is impossible to understand what it being said, though the reader is certainly left unsettled about the notion of proof. As David Stove explains, Lakatos puts quotation marks around words with success grammar like "show," "facts," "discovered," and "proof" so often that it is impossible for the reader to know when, or if, he thinks knowledge ever advances. Stove writes of the use of the word "proof" in Lakatos's book:

> In the book it is subjected countless times to neutralisation or suspension of its success-grammar by quotation marks. Often, of course, equally often, Lakatos uses the word without quotation marks. But what rule he goes by, if he goes by any rule, in deciding when to put quotation marks around "proof" and when to leave them off, it is quite impossible for the reader of that book to dis-

cover. Nor does the reader know what meaning the writer means to leave in this success-word. He knows that the implication of success is often taken out of it; or rather, he knows that on any given occurrence of the word in quotation marks, this implication may have been taken out of it. But what meaning has on those occasions been left in it, he is entirely in the dark.[3]

It can hardly have seemed likely, in the glorious days when Euclid finally delivered the manuscript of the *Elements*, and Plato required the study of geometry of entrants to the Academy, that the distant future would find mathematicians defending their access to the realm of forms by polemics on the misuse of quotation marks. It did not seem any more likely in 1900, but the twentieth century had in store a number of surprises of a similar nature. The forms, unfortunately, cannot defend themselves, as they do not have a causal action on the physical world. Neither ethical nor mathematical truths and ideals can fight tanks or blizzards of quotation marks (though again, neither can they be liquidated by those enemies). They depend on human minds in tune with them to act on their behalf—to implement those ideals and teach them to the next generation. And the necessary defensive action that this implies has to be against whatever attacks have actually appeared, however ridiculous.

Postmodernist Times

Lakatos's work was in the sixties. There is no need to wonder in what direction things have gone since then. Sokal and Bricmont, in their book following up on the Sokal hoax, documented the phenomenon of (especially) French postmodernists persistently choosing specifically mathematical examples to garble. Gödel's theorem, being a subtle and complex result that angels fear to interpret, became such a favorite for postmodernists rushing in that it supplied Sokal and Bricmont a whole chapter of examples.

That chapter opens with a quote from Régis Debray, the French intellectual who joined Che in Bolivia:

> *Ever since Gödel showed that there does not exist a proof of the consistency of Peano's arithmetic that is formalizable within this theory (1931), political scientists had the means for understanding why it was necessary to mummify Lenin and display him to the "accidental" comrades in a mausoleum, at the Centre of the National Community.*[4]

Deleuze, too, had a special taste for pieces of mathematics (or apparent mathematics), such as

> *The respective independence of variables appears in mathematics when one of them is at a higher power than the first. That is why Hegel shows that variability in the function is not confined to values that can be changed ($\frac{2}{3}$ and $\frac{4}{6}$) or are left undetermined ($a = 2b$) but requires one of the variables to be at a higher power ($y^2/x = P$). For it is then that a relation can be directly determined as differential relation dy/dx, in which the only determination of the value of the variables is that of disappearing or being born, even though it is wrested from infinite speeds....*[5]

But actual attacks on mathematics, in the style of the "strong programme" sociologists' attacks on the rationality of science, have been few. The main works in the genre in English are Brian Rotman's 2000 semiotic analysis, *Mathematics as Sign* and Paul Ernest's *Social Constructivism as a Philosophy of Mathematics* (1998). It has to be said that these books are not as far removed from rationality as their titles suggest. They are written in readable English and have arguments that progress according to normal logical principles; perhaps mathematics induces a certain coherence even it its critics. Nevertheless, they do not make much progress in undermining the credibility of mathematics. Rotman's semiotic analysis of mathematical language suffers from the same problem as the efforts of anthropologists to study scientists' "inscriptions" while ignoring what they meant—if the reality of numbers, proofs, symmetries and so on that mathematicians are studying is ignored or "bracketed out," it is impossible to understand what they are doing with their language. Ernest calls attention to the weaknesses of some of the traditional options in the philosophy of mathematics, with some justification, but relies heavily on a brief argument that any reliance on intuition and proof must be unreliable. The argument is entirely contained in the following two paragraphs, which openly express doubts that generally stay hidden under people's intimidation by the prestige of mathematicians. For that reason it is good to have them displayed:

> *It is worth mentioning again the view that some mathematical assumptions are self-evident, that they are given by intuition or some form of immediate access to the (mathematical) objects known. In addition to the problems of subjectivity mentioned above ... there are also those of cultural relativism. Namely, those assumptions that the community of mathematicians regard as self-evident in one era often become the focus of intense scrutiny and doubt in another era (e.g., the axioms of geometry before and after Kant, and the axioms of arithmetic before and after Peano). Self-evidence does not seem to offer a viable basis for justifying the propositions involved, let alone the overall foundations of mathematical knowledge....*
>
> *Therefore, since there is no valid argument for mathematical knowledge other than proof, mathematical knowledge must depend upon assumptions. It follows that these assumptions must have the status of beliefs, not knowledge; must remain open to challenge or doubt; and are eternally corrigible.*
>
> *This is the central argument against the possibility of certain knowledge in mathematics....*[6]

There are two arguments there. The second is that since the regress of reasons ends in axioms or basic propositions, the basic propositions must be dubitable. But labeling axioms as "assumptions" does not make them dubitable. Ernest says that *no matter how strong* his intuition that $2 \times 3 = 3 \times 2$ is, he will regard it as dubitable on the sole ground that it does not follow from anything else. That is a truly heroic level of refusing to know.

The first argument, concerning the alleged changes in axioms from one era to the next, would be a serious one if the facts were correct and axioms did change from one era to another. That is not the case. There are no eras or cultures in which $2 \times 3 = 3 \times 2$ has been denied or doubted. Mathematicians sometimes consider systems similar to—but in certain ways unlike—our number system, but whatever happens in those systems is irrelevant to truths about 2 and 3, just because they are *other* systems and thus 2 and 3 are not in them. The suggested examples of Kant and Peano are wrong. Far from denying or doubting any existing mathematical truths, Kant and Peano were keen to lay down axioms that would produce all the same truths about geometry and arithmetic as previously believed, but to do so with added clarity.

It is true that there is one possible example of doubts about previous axioms, namely the discovery of non-Euclidean geometry. Euclid laid down one set of axioms, but nineteenth-century mathematicians found alternatives. This was a major exhibit of a book once popular in "mathematics for liberal arts" courses, Morris Kline's *Mathematics: The loss of certainty*, which presents a caricature of mathematics during the period 1830–1930 as a discipline lurching from crisis to crisis.[7] It is the sole example of its kind, for one thing, but in any case the example needs to be approached with a great deal of caution. It is true that Euclidean geometry was once thought to be necessarily true of space, and the existence of mathematically possible alternatives showed that the shape of space was an empirical question rather than a purely mathematical one. But that truth is more about the physics of space than it is about mathematics, while an appreciation of the difference between how to measure the shape of space and how to prove geometrical theorems is a matter of philosophy. Euclid's theorems are all still true of the abstract structure, Euclidean geometry. That structure is realized in our space at our scale to a very close approximation, but possibly not exactly.

The writings of the irrationalists about mathematics have had absolutely no impact on mathematicians. Mathematicians have not read one word of any complainers since Lakatos (and they reinterpreted his ideas so as to render them harmless). The production of theorems, as well as the application of mathematics to climate modeling and airline scheduling has proceeded, totally untroubled by any doubts as to the fallibility of proof (but subject to funding, of course).

That's the good news. The bad news is that it is otherwise in mathematics education. There, works like Ernest's *Social Constructivism* are taken very seriously. It may be that the paper "Toward a Feminist Algebra" that was analysed in Gross and Levitt's *Higher Superstition* was somewhat beyond what is typical of the field,[8] but it is not hard to find text still being produced like this extract from *Educational Studies in Mathematics*, 2004:

> *The supposed apolitical nature of mathematics is an institutional frame that functions to sustain specific power structures within schools. This paper disrupts the common assumption that mathematics (as a body of knowledge constructed in situated historical moments) is free from entrenched ideological motives. Using narrative inquiry, the paper examines the ways in which novice mathematics teachers negotiate the intersection of curriculum and institutional politics....*[9]

The corrupters of the corrupters of youth are still hard at it.

CHAPTER 9

The Formal Sciences

It used to be that the classification of sciences was clear. There were natural sciences, and there were social sciences. Then there were mathematics and logic, which might or might not be described as sciences, but seemed to be plainly distinguished from the other sciences by their use of proof instead of experiment and hypothesizing.

The classification was like this:

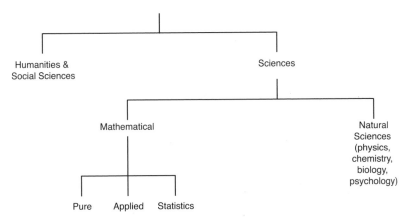

FIGURE 9.1 Traditional classification of the sciences.

That neat picture has been disturbed by the appearance in the last 70 years of a number of new sciences, variously called the "formal" or "mathematical" sciences, or the "sciences of complexity" or "sciences of the artificial."[1] Originally developed mainly as offshoots of various parts of engineering, they have become widely studied in their own right and are hard at work behind the scenes in all complex modern technology. But somehow they have not registered on the radar of the intelligentsia. For that reason we start with a quick survey of what those sciences are and what are a few typical problems considered in each. The classification of sciences is now more like this:

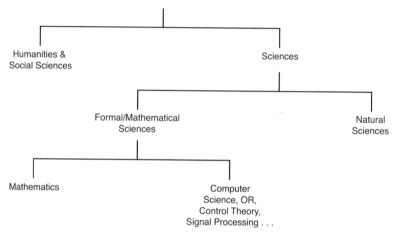

FIGURE 9.2 Current classification of the sciences.

Though the formal sciences have not made much impact on the "learned world" in general, they have two features that should recommend themselves to intellectuals in general— apart from their claims in terms of size and inherently interesting results.

First, they cater well for the word-oriented aspect of spectators of science. If one aim of studying philosophy is to be able to speak plausibly on all subjects, as Descartes says, then the formal sciences can be of assistance. They supply a number of concepts, like "feedback," that permit in-principle explanatory talk about complex phenomena without demanding too much attention to technical detail. It is just this feature of the theory of evolution that has provided a century and a half of delight to intellectual commentators, so the prospects for the formal sciences must be bright.

Secondly, the knowledge in the formal sciences, with its proofs about network flows, proofs of computer program correctness, and the like, gives every appearance of having achieved the *philosophers' stone*—a method of transmuting opinion about the base and contingent beings of this world into necessary knowledge of pure reason. As we will see, that appearance is correct.

Operations Research

While remote antecedents can be found for almost anything, the oldest properly identifiable formal science is *operations research* (OR). Its origin is normally dated to the years just before and during World War II, when multi-disciplinary scientific teams investigated the most efficient patterns of search for U-boats, the optimal size of convoys, and the like.[2] A classic success of early OR came from its analysis of the bombing of U-boats in the Bay of Biscay. U-boats could occasionally be caught on the surface and attacked with a stick of depth charges, but up to 1941 the kill rate in the attacks was only 2–3 percent. Some straightforward analysis of the factors within the control of the attackers— the depth at which the charges exploded, the spacing between charges in the stick, the force of the explosion and the angle of attack—showed that improvements in all of them were worthwhile. In particular, reducing the depth at which the charges exploded from the initial 100 feet appropriate to submerged U-boats to 25 feet made the attacks much more effective. The combined changes resulted in the kill rate improving to 40 percent in 1944.[3]

Typical problems now considered in OR are task scheduling and bin packing. Given a number of factory tasks that take various times, subject to constraints about which must be completed first, which cannot be run simultaneously because they use the same machine, and so on, task scheduling seeks the way to fit them into the shortest total time. Bin packing deals with how to fit a heap of articles of given sizes most efficiently into a number of bins of given capacities.[4] The methods used rely essentially on brute search by computer through all the possibilities, while using mathematical ideas to rule out obviously wrong cases so as to shorten the search.

The Spectrum of the Formal Sciences

Another relatively old formal science is *control theory*, which aims to adapt a system, such as a chemical manufacturing plant, to some desired end, often by comparing actual and desired outputs and reducing the difference between these by changing the settings of the system.[5] To control theory belong two "systems" concepts that have become part of public vocabulary. The first is feedback. The word "feedback" is first recorded in English only in 1920, in an electrical engineering context; outside that area, it appears only from 1943. The second concept is that of "trade-off" ("a balance achieved between two desirable but incompatible features"—*Oxford English Dictionary*). It is first recorded in English in 1961.

There is a not very unified body of techniques that deal with finding and interpreting structure in large amounts of data, called, depending on the context, *descriptive statistics, pattern recognition, data mining, data visualization, signal processing,* or *numerical taxonomy.* The names of products are even more varied: if one purchases a "neural net to predict parolee recidivism" or an "adaptive fuzzy logic classifier," one actually receives an implementation of a pattern-recognition algorithm. Although the science of statistics is rather more than 50 years old, the word usually refers to probabilistic inference from sample to population, rather than the simple finding of patterns in data that is being considered here. When one finds the average or median of a set of figures or divides them into natural clusters, one is not doing anything probabilistic, but merely finding structure in the data. Drawing a bar graph of several years' profits is likewise simply summarizing the data, allowing its structures or patterns to become evident.

These techniques have come into their own with the flood of data now being produced by computerized scanners, weather buoys, surveillance cameras, telescopes, and so on. It is easy to collect and store data, not so easy to "drink from the fire hose of data"[6] and find meaningful structure in it in real time. The classic data-mining story is of Walmart in the mid 1990s trawling through its supermarket-scanning data to find that customers often bought beer and diapers together, and making money by stacking the two together. The story appears to be an urban myth,[7] but it does illustrate well a multitude of real cases where data sets with a huge number of records and many fields in each record are processed to find unexpected patterns useful for prediction.[8]

Then there are several sciences that study flows—of traffic, customers, information, or just flows in the abstract. Where will there be bottlenecks in traffic flow, and what addition of new links would relieve them? Such questions are studied with mathematical analysis and computer modeling in *network analysis*, and there are obvious applications to telecommunications networks.[9] Suppose customers arrive at a counter at random times, but at an average rate of one per minute. If the serving staff can process them at only one per minute, a long queue will form for much of the time. It is found that to keep the queue to a reasonable length most of the time, the capacity of service needs to be about one-and-a-half customers per minute. This is a result in *queueing theory*, a discipline widely applied in telecommunications, since telephone calls also arrive at random times but with predictable average rates.[10] The famous work of Claude

Shannon in *information theory* drew attention to the problem of measuring the amount of information in a flow of 0s and 1s.[11] A sub-branch is the theory of *data compression*: Most messages have many redundancies in them, in that commonly occurring parts (like the word *the* in English text) can be replaced by a single symbol, plus the instruction to replace this symbol with *the* upon decompression. That allows the message to be stored and transmitted more efficiently. There are applications to the DNA "code," with many recent developments on how to deal with the special nature of genomic information.[12]

The concept of expected payoff of different possible strategies for various actors in either a competitive or cooperative environment allows analysis of systems whose dynamics depend on the interactions of such decisions. This is *game theory*. Such systems include business negotiations and competition,[13] animals preparing to fight,[14] and stock-market trading.

More recently there have emerged some overlapping sciences variously known as the *theory of self-organizing systems*, the *theory of cellular automata, artificial life, non-equilibrium thermodynamics*, and *mathematical ecology*. They all deal with how small-scale interactions in large systems create global patterns of organization. As an example, the paradigm of cellular automata is John Conway's Game of Life. On an indefinitely large grid of cells, a few cells are initially chosen as "live." The board then evolves according to these rules for updating:

Death by overcrowding: if 4 or more of the 8 cells surrounding a live cell are live, the cell "dies."
Death by exposure: if none or only one of the 8 cells surrounding a live cell is live, it dies.
Survival: a live cell with exactly 2 or 3 live neighbors survives.
Birth: a dead cell becomes live if exactly 3 of its 8 neighbors are alive.
(Updates occur simultaneously at each time step.)

The remarkable thing is that certain initial configurations lead to complicated and unexpected developing patterns, such as shapes that, after a certain number of "generations," have produced several copies of themselves.[15]

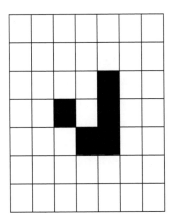

FIGURE 9.3   The "glider" in the Game of Life: After 4 time steps, the same pattern reappears, but shifted.[16]

The study of systems of interacting predators and prey in *mathematical ecology* likewise involves the prediction and explanation of global phenomena from local ones. As prey increase, so do predators, though more slowly. Then if the prey decrease, hordes of hungry predators can nearly wipe them out, leading to the near-extinction of the predators too; then the prey can slowly revive. The discovery of chaotic patterns in the cycles of predators and prey was one of the early discoveries of *chaos theory*, the science of how systems developing in accordance with simple laws can produce random (in the sense of patternless) outcomes.[17] There has been, of course, much resulting speculation about evolution, the origin of the universe, learning in the brain, and so on,[18] some of which will doubtless amount to something someday.

Most of the formal sciences use computers and mathematical modeling in one way or another. Indeed, the advent of the computer has been one of the main factors in the success of these subjects, in allowing results to be obtained in large-scale cases where hand computation is not feasible. But over and above the applications of computing in each science and the engineering of hardware, there exists a *theoretical computer science*. One of its branches is *computational complexity theory*. Typically, one wants to measure the intrinsic complexity of a computational problem in terms of the number of simple operations (additions, comparisons of single digits) needed to solve it. Since computation time is proportional to the number of simple operations, this will show whether it is realistic to solve the problem by computer. For example, the addition of two n-digit numbers (with the usual school algorithm) requires between n and $2n$ single-digit additions. The exact number depends on how many carries there are, as illustrated in the following example, where $n = 4$ and there are three carries:

$$\begin{array}{r} 1\,0\,3\,6 \\ 3\,9\,8\,7 \\ \hline 5\,0\,2\,3 \end{array}$$

This requires seven single-digit additions. Thus, as n grows, the amount of computation needed grows linearly with n, being bounded by $2n$. By contrast, the traveling salesman problem (to find the shortest route that visits n cities once each, given the distances between the cities) demands an amount of computation that grows exponentially with n (at least, that is strongly believed, though not proved). This problem of "combinatorial explosion" makes the traveling salesman problem infeasible for large n (in practice, for n larger than about 40).[19]

As it has proved so difficult to quantify complexity of any kind, computational complexity theory has been studied closely to see if it might be a model for measuring the complexity of things other than algorithms, such as machines or biological organisms. It seems not.

Other issues studied in theoretical computer science include formal specification (to describe exactly what a program is intended to do before it is written), and the effects of a modular or "structured programming" design of programs, which is intended to make understanding and modifying them easier and safer. There is also the discipline of *program verification,* or proof of the correctness of computer programs. (More on this later.)

Usually included in computer science is *artificial intelligence* (AI). We discussed in chapter 6 its success or otherwise in imitating natural intelligence, but in its implementation aspects, the core of AI consists of a combination of computer science and operations-research techniques. To play chess by computer, for example, one employs guided search through the space of all possible moves and counter-moves from a given position. Complexity theory reveals that the space of *all* possible moves is far too big to search, so one observes human players to extract "heuristics," that is, programmable strategies for deciding which of the possible moves are most worth searching next.[20] AI might seem to contradict the assertion that there has been little philosophical interest in any of the formal sciences. It is true, of course, that the philosophy of mind has given much attention to AI, but only for its usefulness as a model of mental workings. True AI workers, on the contrary, tend to be embarrassed by the connection with the mind, and prefer to re-badge their product as "expert systems" or "adaptive information processing." The reason is that computer scientists view AI as an independent discipline concerned with guided search through trees of possibilities, which can only be harmed in the marketplace by unfulfillable claims about imitating the human mind.

Real Certainty: Program Verification

The greatest philosophical interest in the formal sciences is their promise of necessary, provable knowledge that is at the same time about the real world, not just some Platonic or abstract idealization of it. As we saw in chapter 7, mathematics promises that as well, when seen rightly, but various false philosophies of mathematics have obscured that fact. The formal sciences appear to be more directly about the world than mathematics is, so there is some hope that the knowledge in them can dodge the array of defense mechanisms that the underminers of knowledge have ready for mathematical knowledge proper.

The late 1960s were the years of the "software crisis," when it was realized that creating large programs free of bugs was much harder than had been thought. It was agreed that in most cases the fault lay in mistakes in the logical structure of the programs: there were unnoticed interactions between different parts of the "spaghetti"-style code of the time, or there were possible cases not covered.[21] One remedy suggested was that, since a computer program is a sequence of logical steps like a mathematical argument, it could be proved to be correct. The "program verification" project has had a certain amount of success in making software error-free, mainly, it appears, by encouraging the writing of programs whose logical structure is clear enough to allow proofs of their correctness to be written. A great deal of time and money is invested in this activity. The question is, does the proof guarantee the correctness of the actual physical program fed into the computer, or only of an abstraction of the program? C. A. R. Hoare, a leader in the field, made strong claims:

> *Computer programming is an exact science, in that all the properties of a program and all the consequences of executing it can, in principle, be found out from the text of the program itself by means of purely deductive reasoning.*[22]

Some other authors explain the difference between software engineering and traditional engineering with physical components:

> *By contrast [to hardware], a computer program is built from ideal mathematical objects whose behaviour is defined, not modelled*

approximately, by abstract rules. When an if-statement follows a while-statement, there is no need to study whether the if-statement will draw power from the while-statement and thereby distort its output, or whether it could overstress the while-statement and make it fail.[23]

The philosopher James Fetzer,[24] however, argued that the program verification project was impossible in principle. Published not in the obscurity of a philosophical journal, but in the prestigious *Communications of the Association for Computing Machinery*, his attack had effect, as it was suspected of threatening the livelihood of thousands. Fetzer's argument relies wholly on the gap between abstraction and reality, in the same way as Einstein's claim that mathematical models do not directly refer to reality:

These limitations arise from the character of computers as complex causal systems whose behaviour, in principle, can only be known with the uncertainty that attends empirical knowledge as opposed to the certainty that attends specific kinds of mathematical demonstrations. For when the domain of entities that is thereby described consists of purely abstract entities, conclusive absolute verifications are possible; but when the domain of entities that is thereby described consists of non-abstract physical entities ... only inconclusive relative verifications are possible.[25]

It has been subsequently pointed out that to predict what an actual program does on an actual computer, one needs to model not only the program and the hardware, but also the environment, including, for example, the skills of the operator.[26] And there can be changes in the hardware and environment between the time of the proof and the time of operation.[27] In addition, the program runs on top of a complex operating system, which is known to contain bugs. Plainly, certainty is not attainable about any of those matters.

But there is some mismatch between those undoubtedly true considerations and what was being claimed. Aside from a little inadvised hype, the advocates of proofs of correctness had admitted that such proofs could not detect, for example, typos.[28] And on examination, the entities Hoare had claimed to have certainty about were, while real, not unsurveyable systems including machines and users, but written programs.[29] That is, they are the same kind of things as published mathematical proofs.

If a mathematician says, in support of his assertion, "My proof is published on page X of volume Y of *Inventiones Mathematicae*," one does not normally say—even a philosopher does not normally say[30]—"Your assertion is attended with uncertainty because there may be typos in the proof," or "Perhaps the Deceitful Demon is causing me to misremember earlier steps as I read later ones." The reason is that what the mathematician is offering is not, in the first instance, absolute certainty in principle, but necessity. This is how his assertion differs from one made by a physicist. A proof offers a necessary connection between premises and conclusion. One may extract practical certainty from this, given the practical certainty of normal sense perception, but that is a separate step. That is, the certainty offered by mathematics does depend on a normal anti-scepticism about the senses, but removes, through proof, the further source of uncertainty found in the physical and social sciences that arises from the uncertainty of inductive reasoning and of theorizing. Assertions in physics about a particular case have two types of uncertainty: those arising from the measurement and observation needed to check that the theory applies to the case, and those of the theory itself. Mathematical proof has only the first.

It is the same with programs. While there is a considerable certainty gap between reasoning and the effect of an actually executed computer program, there is no such gap in the case Hoare considered, the unexecuted program. A proof is a sequence of steps that exhibits the logical connection between formulas, and is checkable by humans (if it is short enough). Likewise a computer program is a logical sequence of instructions, the logical connections among which are checkable by humans (if there are not too many).

One feature of programs that is inessential to this reply is that they are textual. Although Hoare and the other defenders of program verification began by emphasizing the difference between software engineering and other kinds of engineering, it is possible for hardware as well as software to realize structure to which proof applies. On this view, machines too could be proved to agree with their specifications. Again, it was admitted that there was a theoretical possibility of a perceptual mistake, but this was regarded as trivial, and it was suggested that the safety of, say, a (physically installed) railway signaling system could be assured by proofs that it would never allow two trains on the same track, no matter what failures occurred.[31] The advertisement that said, "VIPER is the first commercially available microprocessor with both a formal specification and a proof that the chip conforms to it," was felt by the experts to be a danger to the gullible public, but not impossible in principle.[32] An aggrieved purchaser began legal action on the grounds that the proof was not complete, but the bankruptcy of the plaintiff unfortunately prevented this interesting philosophical debate from being pursued in court.[33]

Real Certainty: The Other Formal Sciences
The following features of the program-verification example carry over to reasoning in all the formal sciences:

- There are connections between the parts of the system being studied, which can be reasoned about in purely logical terms.
- That complexity is, in small cases, surveyable. That is, one can have practical certainty by direct observation of the

local structure. Any uncertainty is limited to the mere theoretical uncertainty one has about even the best sense knowledge.

- Hence the necessity in the connections between parts translates into practical certainty about the system.
- Computer checking can extend the practical certainty to much larger cases.

Let us recall the classical example of network topology from the last chapter, Euler's proof that it is impossible to walk over all the bridges of Königsberg once and once only. In retrospect the result is the first discovery, in mathematics, of the kind of structure typical of the formal sciences.

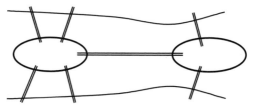

FIGURE 9.4 The Königsberg bridges again.

The first three of the above points are obvious in this example: one perceives all the (relevant) structure and can write it down so as to reason formally about it. The proof demonstrates an impossibility about an actual physical thing that results from that thing's structure. No idealization is needed to obtain a mathematical result.

It is irrelevant that bridges are stone or steel while mathematical proofs and computer programs are logical, or textual. Structured entities in the other formal sciences may be electronic, biochemical, mental, astral, legal, flesh, fish, or fowl. Proofs work by following structure, which may be implemented in steel as well as in text or in abstractions like numbers.

Experiment in the Formal Sciences

Real certainty for armchair work—surely that is too rosy a picture of the formal sciences? If it were right, it ought to be possible to issue real-world predictions by computer without needing to do any experiments. Anyone who has worked in applied mathematics and the prediction of how complex systems behave knows it is rarely like that. It is well known that fitting a realistic mathematical model to actual data is in general difficult. Sometimes, as in meteorology and macroeconomics, it is close to impossible.

To explain when experiment and fitting to data are necessary, one must return to the gap Fetzer insisted on between the abstract model and the real world. Everyone agrees that formal work can proceed with the usual necessity of mathematics, provided one keeps within the model. The important point is that there is wide variability in the certainty as to whether the real world has the structure described by the model. The model-reality gap may be wide or narrow. The word *model* directs attention to cases where fitting is difficult, by the implied suggestion that there may be many models and that the choice among them is difficult. The extreme case is stock-market prediction, where there are plenty of models, but nearly total uncertainty as to which if any fit the data. Any case where an underlying structure has to be inferred from insufficient data will be like that to a greater or lesser extent. The examples above were chosen near the opposite extreme, even, so it was argued, to the extent that there was no gap at all. What structure a system of bridges or a computer program has is open to perceptual inspection, with the practical certainty that attends unimpeded sense perception. So all the hard work is in the mathematics, and the results are directly applicable (again, with practical certainty). Examples like the statistical mechanics of gases fall somewhere in between, but still closer to the formal end. Whether the kinetic theory of gases is true is a contingent fact, not easily established. But it is in fact true, and the way that temperature arises from the random motion of gas particles is a matter of necessity. Though it is harder than in the case of Euler's bridges to determine if things have the properties the model attributes to them, there is real, provable necessity in the *connections* of the properties.

CHAPTER 10

Probabilities and Risks

An understanding of probability is essential to grasping what science knows. As we saw in chapter 1, evidence for scientific truths is often inductive and hence not absolutely certain—sometimes quite uncertain, especially in the early stages of investigation. And some of the material on which science works is inherently variable or stochastic. Thus most of the truths of science can only be known probabilistically—even if sometimes with a very high degree of probability.

Kinds of Probability

Yet probability is an inherently confusing subject. A principal problem is that the word refers to two quite different concepts, and failure to keep them separate has caused endless trouble.[1] First, there is *factual* or *stochastic* or *aleatory* probability, which deals with chance set-ups (dice-throwing, coin-tossing) that produce characteristic random or patternless sequences of outcomes. Almost always, in a long sequence of coin tosses, there are about half heads and half tails, but the order of heads and tails does not follow any pattern. Here is an example of a series of heads and tails from throwing a real coin:

THHHTHHHHHTTHHHTHTHHHHTTTTHHHTHHHTTH

Plainly, the lack of pattern in the outcomes is an objective matter of the way the world is and has nothing to do with any mentalistic or psychological or logical concepts such as uncertainty or predictability. (Though patternlessness may *imply* unpredictability, it is patternlessness that comes first.) There are objective tests of randomness for long sequences. For example, in a random sequence of Hs and Ts, there should be about half Hs and half Ts, and about one quarter of the pairs should be each of HH, HT, TH and TT, and so on. Deterministically generated sequences like the digits of π or the "random numbers" generated by computer can possess the same kind of patternlessness, so it is not a property necessarily associated with anything stochastic.

Notice, too, that the phenomenon of patternlessness in a series of coin tosses does not disappear as we learn more about the microphysics of the situation. It may be that dice throws are fully deterministic, and that if we could measure the initial positions and strength of throw of the coins we could predict the outcome of each throw. Even if we did that with each throw, we would not have explained why about half the throws were heads and half tails, nor why they came up in a patternless sequence. Explaining that needs reference to physical probabilities, whatever account may be given of those.

R. A. FISHER'S TABLE OF RANDOM NUMBERS, 1938

Shortly after our work had been completed there appeared the tables of Professor R. A. Fisher and Dr. F. Yates (1938) which include a table of 15,000 Random Sampling Numbers. This table was compiled from among the 15th-19th digits of A. J. Thompson's *Loga-*

rithmica Britannica.... It includes a careful and elaborate selection of groups of digits by a "random" process involving two packs of cards....

Fisher and Yates applied to their tables certain tests, of which the first is our frequency test. They found an excess of the digits 3, 6, 9, the frequencies being

Digit	0	1	2	3	4	5	6	7	8	9
Frequency	1,493	1,441	1,461	1,552	1,494	1,454	1,613	1,491	1,482	1,519

This gives a value of χ^2 equal to 15.63 with $P = 0.075$. Feeling that this was on the low side, they removed 50 of the 6's "strictly at random" and replaced them by the other 9 digits "selected at random"....

A procedure of this kind may cause others, as it did us, some misgiving....

Maurice G. Kendall and B. Babington Smith, "Second Paper on Random Sampling Numbers," supplement to *Journal of the Royal Statistical Society* 6 (1939): 51–61.

It would not be true to say, though, that probability *just is* the relative frequency of outcomes. The probability of heads is a property of the coin and its throwing mechanism, the *cause* of the relative frequency of outcomes. There is always a possibility of mismatch between the probability of heads and the actual frequency of heads in the throws. Indeed, there is almost always some mismatch: a perfectly fair coin thrown 1,000 times is very unlikely to produce exactly 500 heads. A perfectly fair coin thrown 1,001 times cannot possibly come up exactly half heads, since 1,001 is an odd number. Gross mismatches between probability and outcomes are also possible: one can throw a fair coin indefinitely and always throw heads (though it is not likely). That is what makes factual probabilities *probabilities*—they can always fail to be realized. Thus a factual probability (tendency, bias, propensity) must be regarded as a hidden cause in a repeatable mechanism, which normally will be, but may not be, realized when the mechanism is repeatedly run.

In contrast to factual probability, there is another kind, *logical* or *epistemic* probability, which is concerned with how well-supported a conclusion is by a body of evidence. A concept of logical probability is employed when one says that, on present evidence, the steady-state theory of the universe is less probable than the Big Bang theory, or that the guilt of an accused is "proved beyond reasonable doubt," though not absolutely certain. How probable a hypothesis is, on given evidence, determines the degree of belief it is rational to have in that hypothesis, if that is all the evidence one has that is relevant to it. We saw some examples in chapter 1, with induction and inference to the best explanation. The concept is crucial to theory evaluation in science, as emphasized by the objective Bayesian view of the nature of evidence.

By way of clarification, here is a map of the territory covered by the word *probability*.

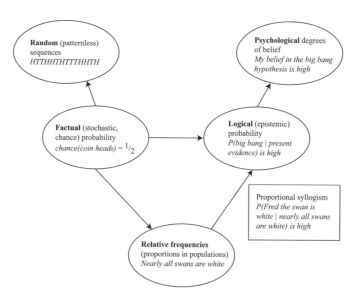

FIGURE 10.1 The kinds of probability.

Logical probabilities cannot be regarded as the same as psychological degrees of belief: degrees of belief *ought* to conform to probabilities—what is really likely on the evidence—but, people being what they are, often do not. The famous psychological results of Kahneman and Tversky showed that humans' deductive—as much as non-deductive—reasoning sometimes fails to conform to sound principles—for example, by being sensitive to the order of presentation of evidence.[2] Failing to distinguish between the logic of science and actual degrees of belief would lead to the philosophy of science being replaced by its history and sociology, a failing whose widespread effects in post-Kuhnian philosophy of science are only too evident, as we saw in chapter 3.

Logical probability cannot be reduced to factual probability. There is no chance, propensity, or frequency involved in, say, whether design arguments for the existence of God are convincing. There is no ensemble of universes throwing themselves, with some relative frequency of divinity in them, but only the question of the conclusiveness of the non-deductive inference from apparent design to an intelligent designer. Conversely, factual probabilities cannot be about knowledge, belief, or logic: the characteristic patternlessness observed in the sequences of dice throws is a physical fact requiring a physical cause.

While factual probabilities, being either proportions or tendencies to produce a definite proportion in outcomes, are inherently numerical, it is not obvious whether logical probabilities should be assigned numbers. Sometimes it is natural to assign a number, as in simple cases of the "proportional syllogism": the probability that an unknown star is dwarf, given just the evidence that 32 percent of stars are dwarf, is naturally assigned a number, namely 0.32. But in more complex cases where the evidence is itself imprecise or bears only remotely on the conclusion, it is harder to say. The legal system has strongly resisted assigning even an approximate number to the standard of "proof beyond a reasonable doubt."[3] Keynes, whose *Treatise on Probability* of 1920 was the original classic of logical probability, believed it was sometimes impossible in principle.

WHAT LEVEL OF PROBABILITY IS "PROOF BEYOND REASONABLE DOUBT"?

In a trial for murder in Victoria a few years ago the jury asked the judge what degree of probability (in terms of percentages) they should require in order to convict of murder. The judge wisely and properly declined to tell them, saying that they had to be satisfied beyond reasonable doubt and that he could give them no further direction. According to a member of the jury, some thought the standard should be set at around 98 percent and some thought it should be 100 percent, and as the jury had to be unanimous ... they decided on a verdict of manslaughter, of which (if murder was rejected) he was clearly guilty....

The most that one can safely say is that the doubt the jury entertains must be a reasonable one and not merely a fanciful one. Lord Denning in *Miller* v. *Minister of Pensions* attempted a more explicit statement: "If the evidence is so strong against a man as to leave only a remote possibility in his favour, which can be dismissed with the sentence 'of course it is possible, but not in the least probable,' the case is proved beyond reasonable doubt, but nothing short of that will suffice."

In the early 1950s Lord Goddard seems to have been assailed by doubts as to the ability of juries to understand the time-honored formula, and a "Practice Note" was issued which said: "It would be better if this time-hallowed expression were no longer used and juries were told that it is for the prosecution to prove the case and that the jury, before convicting, must be satisfied of the prisoner's guilt and feel sure of it. It is to be hoped that this direction, or one in much the same terms, will be generally adopted in future."

The result was unfortunate. Judges, in response to this encouragement, began to invent new ways of expressing the standard of proof. 1961 was a particularly bad year. In no less than seven cases the Court of Criminal Appeal expressed disapproval of the words used by the trial judge.... Among the expressions used were "reasonably sure," "pretty certain," and "pretty sure."

Richard Eggleston, *Evidence, Proof and Probability,* 2nd ed. (London: Weidenfeld and Nicolson, 1983), 114–6.

Of course there are connections between the two kinds of probability, distinct though they are. There are also connections between probabilities and relative frequencies. The most important connection is the proportional syllogism, which states the logical probability of an outcome when the evidence is a relative frequency. A second connecting principle is what the philosopher David Lewis called the Principal Principle, which does the same when the evidence is a statement of factual probability. Some principle is needed to connect any such physical quantity as a factual probability with one's rational degree of belief. If, as a matter of physics, the chance of a die coming up 6 is one-sixth, what should one rationally believe about the next throw of the die? The physics may cause the die to come up 6 about one-sixth of the time (mostly), but what one's degree of belief ought to be is a matter of logic, not physics. The Principal Principle says that on this evidence one's degree of belief in the proposition that the next throw will land 6 ought to be one-sixth.[4] That is a statement of logical probability, and one indispensable for making inference in stochastic situations.

It is the departures from these principles allowed by theorists of the sociology of science and the like that place them in the camp of irrationalism.

Probabilistic Reasoning in Experimental Mathematics

The logical character of probabilistic reasoning is best shown by its applicability in mathematics. The subject matter of mathematics is necessary, so there can be no relevance in considerations about laws of nature, the uncertainty of observations, factual probabilities, or other such things that complicate probabilistic reasoning in natural science. In a mathematical case of a conjecture supported by numerical evidence, there is only the evidence, the conjecture and the logical relation between them.

Since mathematics is a science that describes a normal subject matter, as suggested in chapter 7, it is a scientific study of a world "out there." So, in addition to methods special to mathematics such as proof, ordinary scientific methods such as experiment, conjecture, and the confirmation of theories by observations ought to work in mathematics just as well as in science.

Probabilistic reasoning is found in mathematics in those areas where mathematicians consider propositions that are not yet proved. These are of two kinds. First there are those that any working mathematician deals with in his preliminary work before finding the proofs he hopes to publish, or indeed before finding the theorems he hopes to prove. The second kind are the long-standing conjectures that have been written about by many mathematicians but that have resisted proof.[5]

A particularly straightforward and illuminating example of probabilistic reasoning, in fact of simple induction, in mathematics, is:

> The first million digits of π are random
> So, the second million digits of π are random

The number π has the decimal expansion

3.14159265358979323846264338327950288419716939937510582097494459230781640628620899862803482534211706798214808651328230664709384460955058223172535940812848111745 02 ...

There is no visible pattern in these digits. They appear "random" in the sense of "patternless." (They are, of course, not random in the sense of "stochastically generated" like the outcomes of dice throws.) That initial intuitive impression can be confirmed by statistical tests that show there are about the same number of each digit, in the long run; that there are about the same number of runs up (like 159) and down (like 653) and so on.[6] The sequence of digits exhibits complete pattern-lessness or randomness. The first million digits have long been calculated (calculations have reached beyond 1 trillion digits). Computer calculations applying statistical tests of randomness can confirm they are random. It can then be argued inductively that the second million digits will likewise exhibit no pattern. This induction is a good one (indeed, everyone believes that the digits of π continue to be random indefinitely, though there is no proof), and there seems to be no reason to distinguish between the reasoning involved here and that used in inductions about flames or ravens. But the digits of π are the same in all possible worlds, whatever natural laws may hold in them or fail to. Any reasoning about π is also rational or otherwise, regardless of any empirical facts about natural laws. Therefore, induction must be rational independently of whether there are natural laws.

In mathematics there can be no confusion over natural laws, the regularity of nature, approximations, propensities, the theory-ladenness of observation, pragmatics, scientific revolutions, the social relations of science, or any other red herrings. We see scientific inference laid completely bare, with only the hypothesis, the evidence, and the logical relations between them.

Statistics

R. A. Fisher, one of the founders of modern statistics (and of eugenics) said in 1952, "I venture to suggest that statistical science is the peculiar aspect of human progress which gives to the twentieth century its special character."[7] That gave his own contribution to civilization considerable prominence, but the statement is by no means a gross exaggeration. A mechanical method of generating well-founded generalizations from data is indeed a goldmine for scientific knowledge.

The most spectacular outcome has been drug trials, which have demonstrated the effectiveness and safety of a huge number of beneficial drugs. They use the simple process of comparing the frequency of successes in trials of the drug with those in a control group that does not take the drug—care being taken to have the two populations comparable, randomly selected, and sufficiently large to allow reliable inference.

Fisher explains the essence of the statistical method with the example of the "Lady Tasting Tea." The problem comes from a real event in the 1920s, when one of the women present at an afternoon tea in Cambridge claimed she could tell the difference between a cup of tea where the milk was poured into the cup first and one where it was poured in last. Fisher used it as a hypothetical problem in his semi-popular book, *The Design of Experiments* (1935). How should an experiment be designed to test the claim? The problem is that it is possible to achieve a correct result by chance, especially if the experiment has only a few cups of tea. For example, if the subject gets two out of two right, that is like guessing two coin tosses correctly, which can happen with probability ¼ and hence is likely enough to have occurred by chance.

Fisher recommends that eight cups of tea be prepared (out of sight of the subject, of course), four of them with the milk first and four with milk the last. They should be presented in randomized order (meaning that actual coin throws or a table of random numbers should be used to decide the order of presentation) and the results recorded. He advises against wasting effort on eliminating slight differences in the amount of milk, the strength and temperature of the tea, etc. The number of different ways in which four of the eight cups should have milk first (e.g. the first, second, fourth, and seventh having milk first) can be calculated to be seventy. So if the lady gets them all right, it is very unlikely she did so by chance. As Fisher explains it, if the "null hypothesis"—that she cannot tell the difference and is guessing at random—were true, then she would hit on the right answer only one time in seventy. So the null hypothesis is refuted at a "significance level" of one in seventy (0.014), implying that we rationally have high confidence in our belief that the lady really can tell the difference.[8]

According to someone who claimed to have been present at the original afternoon tea, the experiment was performed and the lady achieved a perfect score, but her name is not recorded.[9]

Extreme Risks, Lack of Data, and Gut Feelings

The utility of the Bayesian approach to evidence is especially evident in the crucially important field of the evaluation of extreme risks. How can one know the probability of events worse than those that have happened so far, given that in the nature of the case there is no or very little frequency data on which to base the probabilities? The problem arises in many fields, such as estimating the chance of terrorist attacks, of major internal fraud in banks, of asteroid strikes, of invasion of pests through quarantine. In all such cases, one is trying to extrapolate the risks beyond the range of the data available. It is no use complaining about the lack of data and the impossibility of calculating precise probabilities: one has to work out which possibilities of disaster have non-negligible risks, so as to take precautions against them.

The essence of the problem of forced decision-making in the face of grossly inadequate evidence was understood in a discussion in the ancient Jewish law code, the *Talmud*. Newborn Jewish boys must be circumcised, but they may be excused if there is a health risk. Is the death from circumcision of a previous son evidence of a health risk? Or of two sons? Induction based on tiny sets of data is unreliable, while speculations about possible causes add nothing. A decision has to be reached, and there is a small amount of relevant evidence that cannot be ignored. The *Talmud* says:

It was taught: If she circumcised her first child, and he died, and a second one also died, she must not circumcise her third child; thus Rabbi. R. Shimon ben Gamaliel, however, said: She circumcises the third, but must not circumcise the fourth child.... It once happened with four sisters at Sepphoris that when the first had circumcised her child he died; when the second [circumcised her child] he also died, and when the third, he also died. The fourth came before R. Shimon ben Gamaliel who told her, "You must not circumcise." But is it not possible that if the third sister had come he would also have told her the same?... It is possible that he meant to teach us the following: That sisters also establish a presumption.

Raba said: Now that it has been stated that sisters also establish a presumption, a man should not take a wife either from a family of epileptics, or from a family of lepers. This applies, however, only when the fact had been established by three cases.[10]

There follows a discussion of whether it is safe to marry a woman who has had two husbands die. It is felt that if the deaths were obviously due to some chance event such as falling out of a palm tree, there is no need to worry; if not, there may be some hidden cause in the woman that the prospective husband would do well to take into account.

To make probabilistic inferences where there is little data and many marginally relevant items of background knowledge, it is usually necessary to fall back on the best (so far) machine for complex reasoning, the untutored human brain, with its intuitions and analogies. Fortunately its performance is excellent in many circumstances. In most of our decisions in life, from where to move to catch a ball to whom to marry, our initial intuitive "gut feelings" are as good a method as any.[11]

That the vast majority of probabilistic inferences are unconscious is obvious from considering animals. For it is not just the human environment that is uncertain, but the animal one in general. To find a mechanism capable of performing probabilistic inference (as distinct from talking about it), one need look no further than the brain of the rat, which generates behavior acutely sensitive to small changes in the probability of the results of that behavior. Naturally so, since the life of animals is a constant balance between coping adequately with risk and dying. Foraging, fighting, and fleeing are activities in which animal risk evaluations are especially evident. The combining of uncertain information from many sources is of the essence of brainpower in the higher animals.

The human species inherited the mammal brain, with these abilities already loaded and in automatic use. In human life, the only certainties, proverbially, are death and taxes; of these, the time and amount, respectively, are quantities rarely known. The "Iceman" discovered in the Alps in 1991 was certainly one who took a calculated risk. There are psychological studies that show how much of cognition generally is "intuitive statistics."[12] Even such a basic operation as the discrimination of stimuli (in deciding, for example, whether two sounds are the same pitch or not) is a probabilistic process of extracting a signal from a noisy background. And perceiving and remembering both involve unconscious testing of hypotheses on the basis of imperfect correlations. Very nearly all uncertain inference is unconscious, performed at the sub-symbolic level by the neural net architecture of the brain.[13]

Human subjective assessments of risk expressed in words are reasonably accurate in many circumstances. Indeed, in business applications like forecasting of such movable quantities as stock prices, human "judgmental forecasting" is still generally comparable to the best statistical methods (and it is possible to say which statistical methods it resembles). As usual, human infants turn out to be champions. They are fantastically efficient learners from statistical data, with the limits to the powers observed in infants apparently being more due to the difficulty of querying them than any lack in their abilities.[14]

We might still prefer not to rely on instinctive human risk assessment, if we had a choice. But we have no choice.

Bank Operational Risk

The ideal way of evaluating extreme risks is to use a group of human experts, under some external oversight to keep them honest, and assisted by technical methods of making the most of limited data. The most intensive effort to do that has been undertaken by one of the most powerful organs of world government, the Basel II compliance regime in banking. At first glance, there is no such thing as a global government: there is no world parliament, and the United Nations has singularly failed to emerge as any sort of credible representative of a global world order. Yet behind the scenes, international cooperation has resulted in a number of powerful world organizations that regulate particular areas forcefully, from the suppression of piracy on the high seas to Internet security. One of the most powerful is the compliance regime that controls the practice of all major world banks. Called Basel II, it is directed by the Bank for International Settlements in Basel. The bank represents only other banks, but its writ is enforced in all civilized countries by government-backed banking regulators (the Federal Reserve in the U. S., the Bank of England in the U. K.) and compliance with its dictates is almost universal. A principal focus of its rules is risk, in particular the amount of capital that banks should reserve against risk.[15]

A bank's business is to take in deposits and lend them out, while reserving a proportion against risks (of loan default, foreign exchange losses, runs on the bank, and so on). The stability of the banking system depends on banks not responding to short-term "market forces" by undercutting one another for short-term advantage and reserving too little (relative to the real long-term risks of their operations). Hence all sound banks have an interest in powerful oversight of the banking industry.

The main principle of Basel II with regard to risk is that banks should have free rein to use any sophisticated statistical methodology they like to calculate their risks, provided they submit the methods and results for the regulator's approval. (This has proved a major source of lucrative careers for mathematics and statistics graduates.)

Of the risks run by banks, credit risk (such as the risk that borrowers will default on loans) and market risk (of trading in, for example, foreign currencies) are comparatively tractable. There is plenty of data and there are sound statistical methods for making predictions from the data. It is quite otherwise with "operational risk," which is a grab bag of unusual and extreme events ranging from massive internal fraud to tsunamis, typing errors in crucial places, incompetent CEOs, and major technological change. A table shows some of the many ways things can go very wrong for a bank:

TABLE 10.1   Types of operational risk.

Type of Risk	Example	Methodology
Internal fraud and human error	Société Generale rogue trader	Model pooled anonymized data, fraud detection
External fraud	Credit card fraud	Fraud detection analytics
Acute physical hazards	Tsunami, hail	Reinsurers' data + extreme value theory
Long-term physical hazards	Climate change	Climate modeling + work on effects on banking system
Biorisks	SARS, animal plague	Biomedical research + quarantine expertise
Terrorism	Bombing, Internet attack	Intelligence analysis
Financial markets risk	2007 subprime crisis, depression	Macroeconomic modeling, stock market analysis + extreme value theory
Real estate market risk	Home loan book loses value	Real estate market modeling

TABLE 10.1 (continued)

Type of Risk	Example	Methodology
Collapse of individual major partner	Enron	Data mining on company data
Regulatory risk	"Basel III," nationalization, government forces banks to pay universities for graduates	Political analysis
Legal risk	Compensation payouts for misinformed customers	Compensation law and likely changes
Managerial and strategic risk	Payout unwanted CEO, dangerous management decision	Consensus of board of directors
Robbery	Electronic access by thieves	Model pooled data, IT security expertise
Reputational risk	Run on bank, spam deceives customers	Goodwill pricing theory + marketing expertise
New technology risk	Technology allows small players to take bank market share	"Futurology"
Reserve risk	Reserved funds change value	
Interactions of all the above	Depression devalues real estate and reserves	Causal modeling of system interactions

The main difficulty with attaching any meaningful numbers to the probabilities of these events is lack of data. Internal frauds, for example, are rarely reported publicly by individual banks unless they are catastrophic. Most banks therefore have little data on past events of the sort that may affect them severely in the future. There is also the fundamental problem, evident in the financial crises of 2008, that previous data may be from "another part of the woods," when the overall system behaved differently, and hence no longer fully relevant. The paucity of data means that it is essential to combine what data there is with expert opinion.

The Basel II regime allows expert judgment to be used, but only within rigid limits. It is mandated that larger banks at least should quantitatively model the probability of losses of various sizes in each of 56 cells—eight "loss types" (such as external fraud, damage to physical assets) in seven "business lines" (such as retail banking, asset management). An individual bank may have no or very few data points (over say the last five years) in some cells, but hundreds in others. It is also mandated that the loss models should take into account four types of evidence: internal data; relevant external data (that is, aggregated data on other banks, possibly in other countries); scenario analysis ("what-if" analyses conducted by teams of experts on situations of financial stress); and "factors reflecting the business environment and internal control systems." The models are expected to use state-of-the-art statistical methods on the available data, but the results are treated as a starting point for negotiations between a bank's internal experts and the regulator, often assisted by expensive expert consultants who mediate between the two. Negotiations are normally conducted in a cooperative attitude that encourages honesty about any gaps in the data and any doubtful aspects of the modeling assumptions.

The result is an "advocacy" model of evaluating extreme risks that could be adopted for many other cases, including assessments of biosecurity and terrorism risks.[16]

Extreme Value Theory

The state-of-the art statistical method most relevant to extreme risks is Extreme Value Theory (EVT). Basel II insists on it in cells with little data. A purely numerical method of extrapolating beyond data, it was originally applied to problems like predicting a river's once-in-a-hundred-years or once-in-a-thousand-years flood height, given the annual maximum height for a shorter period of, say, 50 years. The problem is to look at the largest few floods and extrapolate beyond them to estimate how likely even larger ones are.[17]

To take a real case: radio tracking of mountain brushtail possums in the Strathbogie Ranges, Victoria, yielded 4,996 records of the distances travelled by possums (37 different individuals) from one night's sleep to the next. Many of the records are zero—the possum returns to the same tree. The most typical values are around 60 meters. But there are a small number of very large values, mostly from adolescent male possums cutting ties with home and moving out. The largest 10 observations (rounded to the nearest meter) are:

1256, 1488, 1723, 2011, 2523, 2587, 5492, 5525, 7024, 7152

For purposes such as modeling the possible spread of a possum-borne disease, it is important to estimate the probabilities of very large values—values beyond the range seen in the actual data. Intuitively, the fact that the largest values are considerably larger than the next largest values suggests that even larger values are not unlikely. The data is said to be "heavy tailed," meaning that the "tails," or large values in the data, do not drop off rapidly but are quite spread out. Fitting a standard EVT model to the possum data yields the conclusion that the probability of a possum traveling over 10,000 meters in one day is considerable: large enough to suggest it was a coincidence that such a value was not observed in the data set.[18]

The exact probabilities calculated by EVT need to be treated with caution, as they are sensitive to the few largest values in the dataset, values which could easily have been different. For that reason, more traditional statisticians used to data-rich methods sometimes look askance at the numbers published by their gung-ho colleagues using EVT, especially when the extrapolation is far beyond the data. But EVT is very useful for distinguishing between reasonable and unreasonable estimates of extreme risks. For example, daily stock market data is heavy-tailed—the more extreme events, especially falls, are much more common than would be expected from fitting a simple bell curve to the whole data set.[19] Even if an EVT model fits the tail data somewhat unconvincingly, it is better than imposing an off-the-shelf model that does not fit the data at all. A bell curve advises unrestrained optimism beyond the range of stock market falls that have actually been observed. That is not a rational risk evaluation.

The Unknown Unknowns

Donald Rumsfeld was widely mocked for his tangled distinction between the "known unknowns" and the "unknown unknowns—the ones we don't know we don't know."[20] But he was right. There is a major difference between knowing the hypotheses that might explain a phenomenon and having reasonable probabilities for each, and not having the true explanation within the range of possibilities under consideration. A classic reminder of what philosophers of science call the "threat of the unknown hypothesis" was the 1975 Brown's Ferry nuclear-reactor accident, one of the worst in a Western country. The near-disaster was a result of a concatenation of causes entirely outside the range of what had been considered possible. It fundamentally changed the concept of fire protection and associated regulatory requirements for nuclear power plants in the U. S.

On March 22, 1975, workers at the Alabama plant were fixing leaks in the cable spreading room, well away from the reactor. They used a candle to test seals for air leaks into the reactor building. The temporary polyurethane foam seal, however, was not fire-resistant. The flame from the candle ignited both the seal and the electrical cables that passed through it. The fire quickly spread through the cabling into the reactor.

By the time firefighters extinguished the fire, it had burned for almost seven hours. More than 1,600 electrical cables were affected, 628 of them important to plant safety. The fire damaged electrical power, control systems, and instrumentation cables and impaired cooling systems for the reactor. The operators in the control room should have been monitoring systems and closing them down, but that was easier said than done as the control room was full of smoke. Workers had to perform emergency repairs on systems needed to shut the reactor down safely. There were also major inadequacies in the human response to the unfolding disaster.[21]

Plainly, statistical methods are not capable of finding the chance of something entirely unexpected happening, since statistics is based on counting events in some pre-defined space of possibilities, a space which is by definition not available for unknown hypotheses.

We will have to rely on human intuition—a very fallible tool, but the only one available.

CHAPTER 11

Are the Social Sciences
Sciences?

There is a certain smugness in standard excuses as to why the social sciences—the studies of mass human action like sociology, economics, history, anthropology, linguistics—do not seem to deliver the same kind of precise and testable laws as the natural sciences. It is often said that those sciences, dealing as they do with human decisions in all their marvelous diversity and unpredictability, can hardly be expected to make solid generalizations and predictions. Are we humans not too, well, human, to be reduced to the rule of dusty formulas and rigid laws? As a hostile review of a recent book on "social physics" puts it, "people are not just particles."[1]

The Predictability of Suicides

The definitive answer to this complaint was given in one of the first serious attempts to apply statistical reasoning to social science, Adolphe Quetelet's *Treatise on Man* of 1835. Quetelet was the inventor of the phrase "the average man," which indicates his perspective on the use of data to create summaries of stable averages in the human sciences. He prints this table of suicides in the Department of the Seine for the years 1817 to 1825:

TABLE 11.1 Suicides in the department of the Seine, 1817–1825.

Years	Total	Submersion	Fire-arms	Asphyxia	Voluntary Falls	Strangulation	Cutting Instruments	Poisoning
1817	352	160	46	35	39	36	23	13
1818	330	131	48	35	40	27	28	21
1819	376	148	59	46	39	44	20	20
1820	325	129	46	39	37	32	28	14
1821	348	127	60	42	33	38	25	23
1822	317	120	48	49	33	21	31	15
1823	390	114	56	61	43	48	47	21
1824	371	115	42	61	47	38	40	28
1825	396	134	56	59	49	40	38	20

Suicides are, individually, as humanly variable as it is possible to be, and as firmly the result of acts of the free will undertaken in varied circumstances. Yet as Quetelet writes, "In all the preceding numbers, one may perceive an alarming concordance between the results of the different years, as they succeed each other. This regularity, in an act which appears so intimately connected with volition, will soon appear before us again in a striking manner, as connected with crime...."[2]

No amount of free will in the individual case prevents the phenomena from being predictable (with probability) in the mass, just as the complete randomness of dice throws is not in any way incompatible with the long-term predictability of their average outcomes. While all the potential suicides *could* decide to act in September and not in October, that possibility is no more likely to occur than a run of many heads when throwing dice. Woody Allen explains it in *Love and Death*. He and Diane Keaton are Russians debating the ethics of a plan to assassinate Napoleon.

Woody: Sonja, I've been thinking about this. It's murder.
Diane: If everybody did this, it'd be a world full of murderers.
Woody: If everybody went to the same restaurant one evening to eat blintzes, there'd be chaos. But they don't.

Then again, free will and unpredictability have no close connection in any case. According to Christian theology, Jesus had free will but could always be predicted to do the right thing. Even in our less dependable social world, a person who can be predicted to behave almost always in certain desirable ways is not condemned as coerced or unfree, but praised as reliable. The predictability in the actions of others that we rely on when driving or shopping is quite sufficient to support the statistical generalizations about human behavior that are the stock-in-trade of the social sciences. When I am anxious as my aircraft taxis for takeoff, I am equally reassured by the smooth hum of the engines and by the gravitas of the pilot's voice, and equally worried about the occasional defectibility of aluminum parts and of human decisions.

And on the other hand, the social sciences have no monopoly on unpredictability. The weather is the paradigm of phenomena difficult to predict, and meteorology is certainly not a social science. As we saw in chapter 7, it is understood to some degree why the weather is unpredictable: fluids produce swirls and eddies whose exact development is sensitive to small changes in initial conditions, while the size of the eddies is large on the human scale, so we are unable to predict well events on the scale relevant to us. It is, therefore, the particular kind of complexity the weather system has that causes its unpredictability. Human systems too have dynamics that can create sensitivity to small initial changes—for example, the herd mentality of the stock market can cause panics and bubbles prompted by little in the way of external shocks. What is important to predictability is the kind of complexity that drives the system, much more than whether or not the individual events making up the system are free. The sometimes complained of "unreasonable ineffectiveness of mathematics in the social sciences" is shared with any sciences where complexity gives rise to unpredictable behavior in the mass.

Traffic

In view of the general pessimism about the possibilities of mathematical and scientific methods in the social sciences, it is worth examining briefly one social science in which they have been successful, the study of road traffic. That is not normally thought of as a social science, but it is, in that how a stream of traffic flows is a collective result of the individual decisions of human drivers—decisions to brake, to change lanes, to choose one or another route, all of them made in response to partial knowledge of the road and imperfect predictions of what other drivers will do. It is true that the environment in which traffic moves is much simpler than the one in which, say, political decisions operate. That is the reason why it is simpler to study and why mathematical models have been more successful there.

Models of traffic typically use some simple assumptions about the behavior of individual cars and try to generate some of the characteristic collective phenomena observed in real traffic, such as jams in heavy freeway traffic that appear to happen spontaneously and randomly and then propagate backward along the freeway.

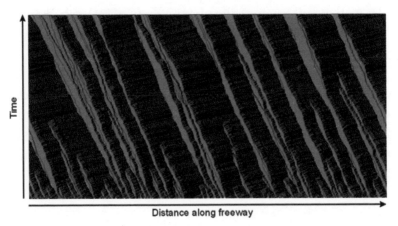

Distance along freeway

FIGURE 11.1   Traffic jams propagate backward on a freeway in the Nagel-Schreckenberg model.

The Nagel-Shreckenberg model is one that is able to simulate such phenomena. It imagines the road divided into cells, each of which may be occupied (by a car) or not. The cars move from cell to cell, with updates at each time step in the computer simulation, and the number of cells moved dependent on the car's speed. Drivers prefer a certain speed, and accelerate to reach it if possible, but decelerate (usually more suddenly) if they find themselves too close to the car ahead (the safe distance being dependent on the car's speed). Some noise is added to the model to simulate imperfections in drivers' estimates and reactions. That model, simple as it is, is enough to generate complex collective behavior qualitatively similar to real traffic. In particular, there is a critical density (depending on speed) at which traffic suddenly becomes congested and subject to random jams. Below that density, traffic flows smoothly and minor random snarls disperse quickly. Above it, traffic jams multiply and the overall traffic flow suddenly slows down.[3]

The model could of course be, and has been, complicated to make it more realistic. However, one point of such "agent-based" models, and of mathematical models in general, is *not* to make them complex, but to find the minimum level of complexity that will generate the phenomenon of interest. Then one knows what is necessary to understanding the phenomenon and what is not.

In the case of traffic, in particular, it becomes clear that, although driven by free decisions of humans, traffic flow only depends on a few features of those decisions, ones that are amenable to mathematical modeling. It is rational to hope that the same is true of economics or history, but there is a long way to go before it will be known if that is so.

Economics, Numbers, and Self-Organization

If economics has not made itself a hard and quantitative science, it is not for want of trying. Business and finance have always been leading consumers of mathematical technology and enthusiastic adopters of new and sophisticated mathematical methods, with the result that economists have rarely shared the innumeracy and math phobia on which some other social scientists pride themselves. In the absence of a Nobel Prize for mathematics, the prize for economics is the one most often awarded to mathematicians.

Mathematical methods have had great success—in certain well-defined areas. Purely statistical methods are good at short-term predictions of sales figures, loan defaults and frauds, for much the same reasons as Quetelet's table gave a reasonable estimate of next year's suicides. If underlying causes stay much the same, statistical methods based on past data produce generally accurate projections and risk estimates.[4]

The trouble starts when underlying conditions change because of some causal shift in the complex system that is the economy. A classic case is the collapse of the inaptly named Long-Term Capital Management in 1998. LTCM, a hedge fund founded in 1994, devised investment strategies based on the deepest available mathematical understanding of the connections between bond rates over time and similar technicalities. Board members Myron Scholes and Robert Merton were awarded the 1997 Nobel Prize in economics for their earlier work on the mathematics of option pricing, confirming that LTCM's mathematical advice was the best money could buy; if those people didn't understand the system, nobody did. For three years they made a lot of money. Following the 1997 Asian financial crisis and the 1998 Russian government default on bonds, bond rates behaved differently from LTCM's predictions and it was revealed where their bets lay. The Federal Reserve organized a $4.6 billion bailout to prevent LTCM's failure from causing a wider collapse of the financial system.[5]

The problem can only be remedied by constructing a causal model of the economy, which would allow the running of what-if scenarios to simulate the effects of changed underlying conditions. It is here that economics has not proved so successful as a science. The prediction of stock prices, economic downturns, interest rate crises, bubbles, and the effects of innovations remains notoriously unreliable. Jokes about the economist who predicted "seven of the last two recessions" are unfortunately to the point.

The fundamental disagreements that still exist as to the nature of the causal structure of economies are illustrated by the debate about Friedrich Hayek and "social justice." Hayek was the "patron saint" of neoliberal economics and one of the principal intellectual forces behind the increasing trends toward deregulation and free markets in the last 30 years. The second volume of his three-volume summation of his later views on society and its organization is entitled *The Mirage of Social Justice* (1975). His thesis that there can be no such thing as social justice, in the sense of ethical principles applicable to overall economic activity, rests on a claim about the causal structure of the economy. Only human actions can be just, he says, so states of affairs such as societal arrangements and economic structures cannot be either just or unjust. For they are just the unintended outcomes of the "self-organization" of society through such means as market forces. His conception of self-organization is similar to that found in traffic models— no one is to blame for snarls in congested traffic. The distribution of wealth resulting from "impersonal" market forces is not the result of the intention of any person or agency. Thinking it is would be to anthropomorphize "society." Therefore, no one is to blame for that distribution, and the concept of justice cannot apply to it.[6]

But this model is plainly excessively simple. It is a matter of the most elementary observation that societies self-organize in a much more hierarchical and intentional way than fluids or ant colonies or traffic. They are much more like a game of a team sport. A game of football can look confused from a distance, and also from the point of view of a camera fixed to the ball. But it is not. The teams have coherence and are directed according to a conscious plan—though a plan that has to keep adjusting to unpredictable elements. The game as played is the outcome of interaction between the planned actions of the team and the cut and thrust of the ball and opposing play. So it is with economic and political ventures. The Great Pyramid, the East India Company, D-Day, Microsoft, and the Catholic Church are great enterprises directed by minds. They are the outcomes of the ability of human institutions to achieve coordinated results by referring decisions to small groups who oversee a complex organization directed to planned results. After the military, capitalist enterprises are among the most successful organizations in taking advantage of such possibilities. Businesses organize complex supply chains and market strategies individually, and the outcomes of those activities are at once large-scale, intended, and partially predictable in their outcomes. They are therefore subject to the demands of justice. To the extent that the distribution of goods is a foreseeable result of those activities—and their point is, after all, to create and distribute certain goods—the distribution of goods is itself subject to the demands of justice. The extent of control or foreseeability of that distribution is not 100 percent, outside a rigid command economy, nor should it be. But neither is it 0 percent or close, as Hayek pretends.

If a model as simplistic as Hayek's can be taken seriously, economics still has a long way to go to reach the status of a real science.

Sociology and the Causation-and-Correlation Problem

In an old joke, researchers train cockroaches to run when they shout, "Go!" They cut off the cockroaches' legs and shout "Go!" The cockroaches don't move. The researchers conclude that cockroaches hear through their legs.

The point of the joke is that inferring causes from correlation is not easy. Characteristic A co-occurs with characteristic B; there is "constant conjunction," in the old philosophers' phrase (or frequent conjunction), but that does not tell us whether A causes B. In general, given that a characteristic A is correlated with a characteristic B (and the correlation is not due to chance), it could be that A causes B, B causes A, or both are the result of some common cause(s) C. Sorting out which is which requires either controlled experiments or extreme care with the naturally occurring data.

The problem affects virtually all sciences that attempt to infer causes from observations of natural systems (as opposed to using carefully controlled experiments with randomized trials). One of the most studied examples comes from epidemiology, which attempts to work out the causes of disease from natural data in human populations. It was believed that taking combined hormone replacement therapy lowered the risk of heart disease, as women taking HRT were observed to have substantially lower than average incidence of heart disease. However, some controlled trials showed the opposite, with women taking HRT having a slightly increased risk of heart disease. On re-examining the data it was found that HRT was more commonly taken by women from higher socioeconomic groups who had lower risk of heart disease in any case, possibly because of better diet and exercise. And doctors may have been more willing to prescribe HRT to healthy women.[7]

Sociology is particularly prone to this problem, because of its vast number of possibly relevant variables and the near-impossibility of carrying out controlled experiments.[8] The newspapers regularly carry stories containing, for example, statistics on the relatively poor health or high crime of people in less well-off areas, with the implication that living in run-down areas causes poor health and high crime rates. It may well do so, but until the figures are corrected to reflect that people in better health can earn more and afford to move to richer and less crime-ridden areas, the raw figures themselves are insufficient to draw a conclusion about the direction of causality.

It cannot really be the case that controlled experiments are informative and uncontrolled observations not. The world cannot "see" the experimenter's will active on it, and there can be natural experiments that could produce the same data as designed experiments. And babies, smart as usual, know how to infer causes from observations (though they do prefer to manipulate causes as well). The problem with both experiments and observations is to explain when and how to infer what causes what, when all one sees in the data is just the pattern of correlations.

This has proved to be an extremely difficult problem. It is far from solved. The leading ideas involve using "Bayes nets," network diagrams with arrows to represent the possible causal influences between variables. Some experts explain the idea in a very simple case:

Consider a simple problem that is far too common for academics who attend many learned conferences. Suppose that I notice that I often cannot sleep when I have been to a party and drunk lots of wine. Partying P and insomnia I covary and so do wine W and insomnia I. There are at least two possibilities about the relations between these variables, which I can represent by two simple graphs: Graph 1 is a chain P → W → I; Graph 2 is a common cause structure I ← P → W. Maybe parties lead me to drink, and wine keeps me up; maybe parties both keep me up and lead me to drink. The covariation among the variables by itself is consistent with both these structures.

You can discriminate between these two graphs by looking at the patterns of conditional probability among the three variables. Suppose you keep track of all the times you drink and party and examine the effects on your insomnia. If Graph 1 is correct, then you should observe that you are more likely to have insomnia when you drink wine, whether or not you party. If instead Graph 2 is correct, then you will observe that, regardless of how much or how little wine you drink, you are only more likely to have insomnia when you go to a party.[9]

That makes the problem sound easy, but it is not. For one thing, there are many more causal structures consistent with the original covariance data, including ones with hidden causes—maybe a hidden mental illness directly causes me to party, drink, and have insomnia. As the very complicated record of psychological experiments on children's causal learning shows, information about temporal order and spatial contiguity of possible causes helps cut down the number of possible causal structures worth considering, but how it all works is far from clear. Then there is the difficult problem of identifying "variables" in the flux of experience in the first place. The babies are ahead of the scientists, again. Attempts to apply Bayes's nets to learning from masses of observations, inferring, for example, causes of disease or crime from databases of personal records, are in their infancy.[10]

History and "Verstehen"

The search for laws of history similar to the laws of physics has a long and dispiriting pedigree. There are no convincing generalities found in the "rise and fall of civilizations" that will enable predictions of the longevity of empires or any similar large-scale trends.[11]

That is not to say, however, that there is anything wrong with historical argument or that it has a logic that is different from the logic of science. History needs to argue from the evidence to the existence of particular facts or summaries of particular facts, such as the existence of the Holocaust. The logic of doing so is the same Bayesian logic we saw in the first chapter, beginning with the confirmation of theories by their consequences.[12] If the Holocaust existed, there should be documents, remains of camps, and millions of people missing. There are, and there is no other remotely credible explanation of that evidence. It is the same kind of reasoning that convinces us that there are stars outside the solar system. Any attempts at a general skepticism about historical evidence will lead to the implausibilities of Richard Whately's 1819 spoof *Historic Doubts Relative to Napoleon Buonaparte*, which imagines what it would be like to argue against the existence of the best-known historical figure of the age.

HISTORIC DOUBTS RELATIVE TO NAPOLEON BUONAPARTE

Whately argues that if the skepticism proposed by Hume and others about reports of miracles were rational, we should on the same principles doubt the existence of Napoleon.

"Let us, if possible, divest ourselves of this superstitious veneration for everything that appears 'in print,' and examine a little more systematically the evidence which is adduced ... what means have the editors of newspapers for giving correct information? We know not, except from their own statements. Besides what is copied from other journals, foreign or British, (which is usually more than three-fourths of the news published,) they profess to refer to the authority of certain 'private correspondents' abroad; who these correspondents are, what means they have of obtaining information, or whether they exist at all, we have no way of ascertaining. We find ourselves in the condition of the Hindoos, who are told by their priests that the earth stands on an elephant, and the elephant on a tortoise; but are left to find out for themselves what the tortoise stands on, or whether it stands on anything at all....

What, then, are we to believe? If we are disposed to credit all that is told us, we must believe in the existence not only of one, but of two or three Buonapartes; if we admit nothing but what is well authenticated, we shall be compelled to doubt of the existence of any....

'But what shall we say to the testimony of those many respectable persons who went to Plymouth on purpose, and saw Buonaparte with their own eyes? Must they not trust their senses?' I would not disparage either the eyesight or the veracity of these gentlemen. I am ready to allow that they went to Plymouth for the purpose of seeing Buonaparte; nay, more, that they actually rowed out into the harbour in a boat, and came alongside of a man-of-war, on whose deck they saw a man in a cocked hat, who, they were told, was

Buonaparte. This is the utmost point to which their testimony goes; how they ascertained that this man in the cocked hat had gone through all the marvellous and romantic adventures with which we have so long been amused, we are not told."

Richard Whately, "Historic Doubts Relative to Napoleon Buonaparte" (1819). Available online: http://en.wikisource.org/wiki/Historic_Doubts_Relative_to_Napoleon_Buonaparte.

There is however one important difference between the human sciences, especially history, and the natural sciences. It is expressed in the German word "Verstehen," literally "understanding," which refers to the possibility of an imaginative insight into the decisions of other humans. Concepts that need to be expressed in German are, in general, dubious, but this is an exception to that generalization. One cannot imagine what it is like to be an alpha particle heading for a slit in a physics experiment (because there is nothing it is like to be an alpha particle), but one can put oneself, though imperfectly, in the place of a general deciding on strategy or a banker deciding on a weighting of an investment portfolio. That gives one an extra mode of understanding of those phenomena of the social sciences that do depend on individual human decisions.

The development in infants of the ability to "read" others' minds is one of the most fascinating branches of developmental psychology. Newborns already react differently to people and objects, but from about the age of one they develop abilities such as pointing that imply they are starting to understand that other people view things the way they do ("To point to something, especially when you point again and again, looking back at the other person's face until he or she also looks at the object, implies that you think, at some level, that the other person should look at the same thing you are looking at."[13]) Eighteen-month-olds can understand that an adult might prefer a different food from themselves. Three-year-olds can understand the holding of different beliefs about the same thing, finding it hysterically funny that an adult coming into a room does not know about an object that they have seen being hidden. Those cognitive abilities are needed to make sense of the social world. They are equally necessary to making initial sense of the world of other humans, whether other tribes or historical actors.

An approach involving Verstehen is sometimes essential to understanding an historical event, even a large-scale one. Take the case of Operation Barbarossa, Hitler's surprise attack on the Soviet Union on June 22, 1941. The largest invasion in history, it resulted in due course in the end of Nazism and the saving of the hard-pressed Western allies, at the cost of some 30 million military and civilian casualties. Certain aspects of it can be explained in terms of the strengths of the opposing forces, but the decisions and beliefs of Hitler and Stalin played an essential role in the course of events. Hitler and Stalin were neither entirely rational nor completely insane, and it is possible to understand, to some degree at least, how they came to believe what they did and how that influenced the course of events.

There was no inevitability to the invasion, much less to the timing of it. While Nazism and Communism were naturally antipathetic, the Molotov-Ribbentrop pact of 1939 had held solid, and the two sides had cooperated, especially in dividing Poland. There was no compelling reason for either side to breach the peace. Hitler had plenty on his hands with an undefeated Britain, something Stalin knew well and that helped convince him Hitler would not invade. Hitler, however, believed in the weakness of the U. S. S. R., a theory confirmed by its poor military performance against Finland; his belief that "we have only to kick in the door and the whole rotten structure will come crashing down"[14] is essential to understanding his decision. Stalin, for his part, matched Hitler in arrogance in choosing to trust Hitler and believe the German disinformation campaign instead of a flood of intelligence on the German buildup in the

East.[15] He therefore avoided serious preparations for an invasion in order not to antagonize Hitler.

The essential role of Verstehen in understanding history in a case like this arises from two factors. The first is that one important way human society organizes itself is hierarchically—organizations form in such a way that strategic decisions are referred up the line to a central leader or small committee. Hitler and Stalin were rightly called dictators because their personal decisions dictated what course whole nations took. Given that, an understanding of why they made those decisions is essential to knowing why events unfolded as they did. How the facts would have looked to an objective observer is one element, as are psychological speculations as to how the dictators' arrogance distorted their evaluation of that evidence. Our imaginative putting of ourselves in their shoes is, though a fallible tool, the best one we have for those purposes.

That is not to say that the method of Verstehen has a different logic from purely empirical inquiry. Methods like the confirmation of theories by their consequences apply just as well to a theory of Stalin's thinking as they do to a theory of Stalin's bodily health. If it turns out that Stalin chose to purge intelligence advisers who told him the truth (as he did) that confirms a theory of his arrogance and imperviousness to evidence.

So the answer to the question, "Are the social sciences sciences?" is, "Yes and no." Their logic of confirmation of theories is the same Bayesian reasoning as is found in science, and theories in the social sciences can be established beyond reasonable doubt. Statistical methods can predict short-term trends from data just as well as in the harder sciences. But finding the appropriate concepts and the causal connections between them is much more difficult, while imaginative insight into the minds of individual actors gives an extra mode of understanding unavailable outside the sciences of the human.

CHAPTER 12

Actually Existing Science
Institutions for Knowing

The Poles, their cynicism finely honed by Cold War realities, used to speak sneeringly of "actually existing socialism." The phrase pointed up the contrast between the ideals of socialism and what had become of its institutions on the ground. In principle it could be the same with science. Science the theory has to be produced by science the practice. Theories may exist in a Platonic realm, or in minds, but practices cost money, so they are subject to the political, ethical, and other causal forces that any real-world practice has to deal with. If postmodernist sociology of science and the conclusions drawn from it are wrong, it is still true that there must be a genuine sociology of the scientific community—and a psychology, history, politics, and ethics of science. Those disciplines should tell us whether science, the enterprise, is or is not well-organized to deliver science, the set of truths. They should provide materials for answering such questions as why modern science arose only in the West and whether on balance science has been of benefit to humankind.

They might like to explain also how it was possible that there sometimes persisted for long periods science-like institutionalized bodies of opinion—alchemy, astrology, phrenology,

197

Lysenkoist biology, Nazi racial "science," and "creation science" —that were not oriented to delivering truths. Medicine up to the eighteenth century was, on a modern understanding, hard put to demonstrate any net benefit to its patients (over and above such Stone Age techniques as setting broken bones), but doctors did not lack title, respect, or income.

Those are vast questions. Here we can only work through a few vignettes to sample what those fields have to offer.

Psychology of Science

The psychology of science ought to be the most basic of the social studies of science, since all science is done by human minds (with only marginal assistance from computers and instruments, themselves under the direction of human minds). Yet it is the least developed of the studies of science, with the first dedicated journal, the *Journal of Psychology of Science and Technology*, beginning publication only in May 2008.

Research in the field has however been able to identify a number of "risk factors" for becoming a successful scientist, in the sense of traits represented more frequently among successful scientists than in comparable populations of non-scientists. It helps to be Western, male, from an immigrant family, Jewish, descended from scientists or engineers, and firstborn, and in personality to be curious, intelligent, arrogant, and driven.[1] For mathematics and physics especially, a touch of autism and introversion does not go astray.[2]

It is also possible to say something about the psychological skills needed for science. Success needs not only persistence and patience, but some more particular mental abilities needed to evaluate evidence judiciously. "The development in scientific thinking believed to occur across the childhood and adolescent years might be characterized as the achievement of increasing cognitive control over the coordination of theory and evidence ... it entails mental operations on entities that are themselves mental operations."[3] One must first separate theory and evidence—something non-scientific thinkers find difficult—and appreciate that theories, including those that one believes or wants to believe or not to believe, have evidence both for and against them. Suspense of belief and commitment when justified are both cognitive and moral skills.

Yet it is hard to extract from these findings much useful advice on how to organize science better or even insights on where science may be going right or wrong. Here as elsewhere, the concept of "risk factors" deliberately runs together causes and symptoms, so one does not know which variables one could manipulate to change outcomes. Despite intensive study, it is still not clear what the roles of nature versus nurture are in the predominance of males at the highest levels of science, especially physics and mathematics. And if some trait or group in the community is underrepresented in science, it is not clear—especially in the absence of an established causal story—what policy advice should follow from that: should one concentrate resources on groups that have shown they can achieve so as to pick more winners, or should one diversify so as not to miss talent in underrepresented groups? It is hard to go beyond stating the obvious—that quality science and mathematics education across the board is essential and that there must be programs to identify and encourage children who show special interest and talent in science.

Sociology of Science: Grants

A genuine sociology of science—using a combination of statistical and Verstehen-based methodologies, like any serious sociology—should have valuable contributions to make in understanding, and possibly improving, how science carries out its task of discovering the truth. Science is a very expensive business, hence subject to social and political pressures and competition for resources that deserve study in their own right.

In simple terms, if scientists were better at sociology, politics, and marketing (as good as lawyers, for example), they might be given more money.

There are three models for distributing money for scientific research: personal patronage, competitive public funding, and patentable commercial research. Some scientists have been exceptionally good at convincing rich patrons of their genius. Galileo had those skills when young, though he lost his touch later in life when dealing with the Pope.

HOW TO WORK FAST ON A PATRON

They worked quickly in those days, but an eye for a grant came in handy as much then as now. In June 1609 Galileo, then Professor of Mathematics at Padua but always hopeful of a move back home to Florence, heard an interesting rumour from one of his dubious intellectual contacts in Venice. It was said that someone in Holland had put lenses in a tube and was able to see things larger at a distance. Galileo did some geometry as to what shaped lenses might do that and over several months of experiment (and learning to polish glass) he achieved a result that magnified 9 times with good clarity. In August he exhibited it to the senators in Venice. The military implications were obvious and they doubled his salary. By the winter he had a version that magnified 20 times, but with rumours all over Europe it was going to be hard to keep ahead of the competition. On Jan 7, 1610 he turned the telescope on Jupiter. It had three unknown stars next to it, in itself not surprising as many new stars were evident. The next night, the tiny stars seemed to have got ahead of Jupiter instead of being left a little behind. Jan 9 was overcast. On Jan 10, the stars were still near Jupiter but in a different position. Galileo realised he had discovered moons, the first celestial discovery in recorded history (except for transient events like comets). A fast enough printing job would ensure his priority, and he had the naming rights. As a result of a previous networking success, he had some years before tutored the young Grand Duke Cosimo de Medici of Florence, before his accession. On Feb 13 he wrote asking for gracious ducal permission to name the new moons the "cosmic" stars, suggesting Cosimo, but adding that "Medicean" might also be a possibility. The Grand Duke's marketing department replied graciously preferring the latter option, and with the necessary last-minute typesetting changes the book announcing the discovery of the Medicean stars appeared on Mar 12. The several pages of dedication to Cosimo include the sentence "Scarcely have the immortal graces of your soul begun to shine forth on earth than bright stars offer themselves in the heavens which, like tongues, will speak of and celebrate your most excellent virtues for all time." There was then an urgent need to fend off a claim by a rival professor of mathematics that it was all made up, but an urgent request to Kepler and a rushed pamphlet from him (though he did not have a telescope powerful enough to see the new stars), sorted out that contretemps. Galileo was appointed grand ducal mathematician and philosopher at Florence on July 10, needless to say at an acceptably huge salary.

Summarized from Atle Næss, *Galileo Galilei: When the World Stood Still*, trans. James Anderson (Berlin: Springer, 2005).

The patronage model is not out of the question even now, when there exist extremely rich individuals who hope to be remembered for something other than just being rich. However, such private initiatives as the Bill and Melinda Gates Foundation's grants for malaria research[4] tend to model themselves more on public grant processes than on individual largesse in the style of Renaissance princes.

Funding by competitive grants (on top of funding of research infrastructure) remains the principal method for funding basic research and certain kinds of applied science, such as long-term military research. A team of researchers makes a pitch to a funding agency, detailing their track record, research plan, and promises of wonderful outcomes. The grant body takes advice from relevant experts and makes a decision about which projects to fund. Funding is for a fixed period, typically three years. The system works adequately and there is no obviously better plan, but it has weaknesses. The experts consulted are probably either friends or rivals of the applicants if they are close enough to the field to be fully informed. Public funding tends to be fickle, both in total quantity and in commitment to particular areas, so that someone who has made the necessary investment of a whole career in one area may be left high and dry, to their personal disadvantage and to the discouragement of young scientists considering their career options.

WHEN THE GRANTS RUN OUT

Dr. Katharine Jefferts Schori gained a PhD in marine biology ("Zoogeography and Systematics of Cephalopods of the Northeastern Pacific Ocean," Oregon State U, 1983) and worked with the National Marine Fisheries Service in Seattle. In 1986 there was a squeeze on the federal research budget. "I couldn't find a job unless I wanted to write grant proposals all the time," she said, "and that wasn't the part of the ocean I enjoyed." She left science and moved to another industry. In 2006 she was elected Presiding Bishop of the United States Episcopal Church, the first female primate in the Anglican communion.

Carol Reeves, "Minister Says Farewell to Corvallis," *Gazette-Times*, January 6, 2001, http://oregonstate.edu/~schorir/mv_religion.html.

From the perspective of scientific progress as a whole, a serious problem with the grants-and-patronage model of funding science is that it does not feed back into science a fair share of the economic benefits arising from it. The computer, the Internet, and e-mail were spinoffs of publicly funded military research. The World Wide Web (in the sense of the system of websites and browsers) was invented by IT staff at CERN Geneva to share physicists' papers.[5] Those fantastically productive ideas were given away absolutely free to the business world, which had consistently failed to innovate in large-scale computing. Those gifts being free helped their quick adoption. Any reasonable estimate of the economic benefits would have been enough to fund basic science research many times over, but business has never made any moves to repay the debt, and there is no system that could require it to do so.

Direct feedback of money from profits—and, hence, a flow of funding to new research—is possible when the research outputs are patentable. This is possible for applied rather than basic research, because the benefits of basic research are hard to attribute and because the legal system permits the patenting of "expressions" of ideas but not of the ideas themselves. Commercial research has proved successful in such areas as pharmaceuticals and biotechnology. Give or take some problems over the skewing of directions of research, such as a preference for research on diseases of the rich like obesity over diseases of the poor like malaria, patents in biology have proved a remarkably effective means of channeling money from sales of beneficial research outcomes back into new research.

Sociology of Science: Refereeing

Perhaps the most pressing sociological issue for the orientation of science toward finding the truth concerns peer review. Anyone with the necessary resources can conduct scientific research, make claims about it on the Web and at press conferences, and submit it to a scientific journal. But it only counts as accredited science if it is accepted by a journal after being approved by the journal's referees. The procedure is that the journal's editor(s), after an initial judgment on whether the paper's topic is "suitable for the journal," sends it for refereeing (normally without the authors' names attached) to a few of the journal's contacts who are considered experts in the area. The referees are expected to judge both the correctness of the paper and its value (significance and originality). In the light of the reports (which are sent anonymously to the authors), the editor rejects, accepts, or "accepts subject to revision." A rejected paper—and a top journal in a field may be accepting only 2 or 3 percent of submitted papers—is typically resubmitted to a less prestigious journal, possibly with changes or possibly not.

Refereeing is plainly necessary and is crucial to the credibility of published scientific results. If a paper has passed scrutiny by genuine experts who have found the time to concentrate on it, it is probably correct. Pseudosciences do not have a serious refereeing process, which is one reason why they remain pseudosciences. The process does work reasonably well in, for example, pure mathematics, where the reasoning is all in the paper and an expert can usually find a substantive error in a short time; the field is also not flooded with too many papers per researcher and per reviewer. It is rare, though not unknown, for pure mathematics papers that are studied later to have errors exposed.[6] In all areas, refereeing acts in the same way as the police hope to do in keeping the peace—by long-range deterrence more than hands-on action. The fear of being refereed anonymously spurs researchers to avoid errors.

But the process is inherently flawed. One obvious problem is that refereeing is unpaid, and the saying, "If you pay peanuts, you get monkeys" is applicable. Journal-publishing companies pay neither authors nor referees (and pay editors very little), while charging those academics and their institutions large fees for subscriptions, so the motivation for busy academics to spend serious effort on refereeing is low. The commitment of academics in general to refereeing can be measured by the time they take to get around to doing it, which can sometimes match that of the proverbial monkeys typing Shakespeare. If the time taken to begin reviewing a paper is large, the time actually spent on it is not—in one study, reviewers spent an average of 2.4 hours per paper.[7] For any given paper, the scientists most capable of refereeing it competently are not only too busy to do so but, as with book reviewing in the humanities, will often find it worthwhile only if the work under review is written by a friend or an enemy (which even under conditions of blind refereeing will generally be obvious). Those conditions are not conducive to objective and thorough reviewing. The editor-in-chief of *The Lancet* wrote, "We portray peer review to the public as a quasi-sacred process that helps to make science our most objective truth teller. But we know that the system of peer review is biased, unjust, unaccountable, incomplete, easily fixed, often insulting, usually ignorant, occasionally foolish, and frequently wrong."[8] He should know.

The lack of resources for refereeing is of particular concern for experimental papers. Referees certainly cannot spare the time to replicate experiments, so they generally have to take the proffered experimental data on trust if it "looks reasonable." That means the process is not well-adapted to discovering fraud; one commentator suggests the record on the detection of fraud shows that "suspicious competitors, aggrieved postgraduate students, incredulous promotions committees and jilted lovers" have often done a better job than referees of identifying anomalies that should be investigated.[9] The process can also allow the submission of results that are really due to chance, according the principle that if one wants results at the 5 percent significance level one should have 20 graduate students repeating the experiment. The refereeing process has come under severe strain as a result of the vast flood of research by pharmaceutical companies, nearly all of which claims to prove the effectiveness of the sponsors' products and where there is little indication of which trials were aborted in the early stages.

Underresourced refereeing is also a problem for research that may look unreasonable because it is groundbreaking although true. A classic case is the discovery of the Belousov-Zhabotinsky reaction, a truly bizarre phenomenon in chemistry in which a solution undergoes spontaneous waves of color changes. Belousov's original paper was rejected on the grounds that such a phenomenon was impossible, and he was never able to publish more than a short abstract in an obscure journal.[10] Admittedly, the story is from the U. S. S. R. in the 1950s, not a period renowned for its openness to new ideas, and it is not clear how present journal editors balance innovation and improbability of claimed results.

An example of the opposite error comes from the "Bible Code" affair, where the editors probably acted reasonably but the innovative result published was wrong. In 1994, three Jewish mathematicians submitted a paper "Equidistant letter sequences in the Book of Genesis" to the respectable journal *Statistical Science*. This paper claimed that if one searched by computer through the sequence of the 304,805 Hebrew characters of the Torah, looking at "words" formed by starting at random points and with fixed intervals between (for example, starting at character 107 and taking every fifth character thereafter 8 times), one would find the names of the Great Rabbis of the Jewish tradition much more often than would be expected by chance.[11] The skeptical editors subjected the paper to an unusual three rounds of peer review. The reviewers were unable to find anything wrong with the statistical analysis. The editors took the courageous decision to publish the paper, with the comment that it was a "puzzle." Controversy developed rapidly, especially with the publication of a popular book, *The Bible Code*, whose predictions based on similar methods included (possibly) the assassination of Yitzhak Rabin the year before it happened. The book reprinted the *Statistical Science* paper, making it probably the most widely disseminated mathematics paper ever.[12] Replies, including one in *Statistical Science,* convincingly suggested that flexibility in the spelling of the Rabbis' names had allowed the original authors extra leeway in fitting to the data that vitiated their analysis.[13] Since that explanation implicitly accused the original authors of dishonesty, acrimony ensued.

The outcome was scientific progress. Nevertheless, a refereed article even in this purely mathematical area was incorrect.

It is also possible for a dissident group of scientists to escape the scrutiny of hostile establishment referees by setting up their own journal, where they can referee one another's papers to their hearts' content. If they can find subscribers, publishers are always looking for new journals.

RANDOMLY-GENERATED PAPER ACCEPTED BY DODGY CONFERENCE

Could the Sokal hoax be perpetrated in reverse, by a completely meaningless scientific paper being accepted by a journal or conference in science? There is a putative example. The randomly-generated paper, "Rooter: a methodology for the typical unification of access points and redundancy" was submitted to the 9th World Multi-Conference on Systemics, Cybernetics and Informatics (Orlando, Florida, July 2005).

The abstract is: "Many physicists would agree that, had it not been for congestion control, the evaluation of web browsers might never had occurred. In fact, few hackers worldwide would disagree with the essential unification of voice-over-IP and public-private key pair. In order to solve this riddle, we confirm that SMPs can be made stochastic, cacheable and interposable."

The conference organizers wrote back that as no reviews had been received, "your paper has been accepted as a non-reviewed paper" for presentation. On being informed of the hoax, an organizer replied, "I am not ashamed of myself but having a huge sadneess."

The event is interesting, but as the conference itself was of dubious standing and the paper was not in fact accepted as refereed, it is far from being a parallel of the Sokal hoax. Outsiders might also wonder if the abstract is especially more meaningless that most of what passes for text about software.

See "SCIgen—An Automatic CS Paper Generator," http://pdos.csail.mit.edu/scigen/#examples.

All in all, it is no surprise if the probability of a refereed paper being simply wrong in its conclusions is substantial. This has been exposed most brutally in law courts where opposing lawyers in product liability compensation cases have had no trouble in producing rival scientific experts, both teams being able to exhibit an impressive array of refereed scientific papers claiming that cause X did (or, respectively, did not) cause adverse effect Y. And as we have seen, the inferring of causality from observational results is a process fraught with difficulty at the best of times.

Controversy has centered on the 1993 *Daubert* decision of the U. S. Supreme Court, which laid down standards for the quality of any scientific research that sought to be accepted in court. The court in effect took a position on the philosophy of science. Previous legal practice had taken a *laissez faire* approach to the tendering of evidence, allowing opposing counsel quite free rein to supply experts on any scientific matters of contention and permitting judges and juries to reach their own conclusions. Concerns were raised by Peter Huber's 1991 book, *Galileo's Revenge: Junk Science in the Courtroom*, which pointed out the possibility that superficially plausible claims by pseudoscientists would fool ignorant judges and juries, and prevail over respectable scientific claims, to the benefit particularly of plaintiffs bringing unjustified claims for compensation. *Daubert* laid down that the testimony of (alleged) scientists should meet certain standards before being admitted in court. In the case the Court was reviewing, Jason Daubert had been born in 1974 with only two fingers on his right hand and without a lower bone on his right arm. His mother had taken the drug Bendectin in pregnancy and sued the manufacturer, Merrell Dow Pharmaceuticals Inc. The issue was whether there was evidence that Bendectin caused birth defects. The manufacturers presented evidence of a review of scientific studies of humans that concluded Bendectin was safe, implying that Jason Daubert was one of the proportion of births with deformities that occur naturally. The plaintiffs also had scientific experts testifying for them, but their evidence consisted of animal and non-peer-reviewed studies that were clearly inferior as science to the evidence of the manufacturers. The Supreme Court held that evidence of such poor scientific quality should not be allowed in court in the first place. They laid down criteria (though not to be used as a "definitive checklist or test") whereby judges could distinguish sound, legally acceptable science from junk. The criteria are:

1. Whether the theory or technique has been or can be tested, i.e. falsified (explicit mention was made of Karl Popper, but "falsification" appeared to be understood in a much looser sense than in Popper's deductivism).
2. Whether the theory or technique has been subjected to peer review and publication.
3. Consideration of the known or potential rate of error of the method used.
4. The existence and maintenance of standards controlling the technique's operation.
5. Whether the scientific community has generally accepted the theory or method.[14]

Criteria 1 and 3 are logical, criteria 2, 4, and 5 are sociological. If used not too rigidly, the test is reasonable as a way of distinguishing obviously respectable science from obvious rubbish, and hence of preventing the latter from wasting the time of the courts. The criteria are meaningful to an intelligent but non-scientific (and non-philosophical) audience. Nevertheless, in any case where the scientific evidence admits of genuine doubt there will be the problem, raised by Justice Rehnquist in dissent in the original case, that judges are constituting themselves gatekeepers of science while lacking scientific expertise. The problem is the same one as continually arises in medical negligence cases: if legal instead of scientific experts are given the last word on decisions that are essentially scientific, there will be an unnecessary number of incompetent decisions made.

Sociology of Science: Attracting the Next Generation

A sociological issue just as important to scientific truth as peer reviewing is the attraction of talented young people into research. Rather than review the findings on the issue, let us simply repeat a joke that highlights one of the most important things to know about how to become a successful scientist.

SUCCESS IN YOUR THESIS: IMPORTANT ADVICE

Once upon a time in the forest, a fox came into a clearing. It saw a rabbit. The rabbit was typing.

The fox said "What are you doing?"

The rabbit said "I'm writing my thesis."

The fox said "Oh really? What's it on?"

The rabbit said "It's entitled 'The Role of the Fox in the Diet of the Rabbit.'"

The fox said "That's ridiculous. It's foxes that eat rabbits." The rabbit said "No, come down the rabbit hole here and I'll show you how it's done."

So they went down the hole.

After a while, the rabbit came out alone, licked its lips, and went back to the computer.

Some time later, a wolf came into the clearing.

"What are you doing?"

"I'm writing my thesis."

"What's it about?"

"It's on 'Wolf Protein as a Source of Rabbit Nutrition.'"

"Don't make me laugh. Wolves eat rabbits, not the other way round."

"No, no, come and I'll show you."

They went down the burrow together. The rabbit came back by itself, burped, and went back to the word processing.

Later a bear came into the clearing.

"What are you doing?"

"I'm writing my thesis."

"Oh yes, what's it on?"

"'Why Rabbits Eat Bears.'"

"Don't be stupid."

"Come and I'll show you."

So they went down the burrow.

Down there, they came into a big room. In one corner, there was a pile of fox bones. In another corner, there was a pile of wolf bones. In a third corner there was a lion picking its teeth.

Moral: It doesn't matter what your thesis is about, or what you use for data or conclusions. *The important thing is to have a good supervisor.*

Internet joke of 1995, slightly adapted.

The Ethics of Science

Of the many ethical issues concerning the uses of science, there are a few especially concerned with its ways of knowing.

There is, unfortunately, a dark side to the obsessive commitment to knowledge that is the hallmark of the best science. The image of the "mad scientist" is not totally wide of the mark. Certain scientists, as might be expected from the overrepresentation of a touch of autism in the scientific personality, have taken their commitment to finding the truth to lengths that have deliberately harmed everyone from their neglected families to the human and animal subjects of their experiments, sometimes even themselves. The Nazi and Japanese medical experiments of the 1940s are well-known.[15] Even after those murderous regimes were eliminated, the mid-century prestige of science led to certain grossly unethical programs of experimentation, such as the

CIA's mind-control experiments on unwitting subjects in the 1950s.[17] Less spectacular, and indeed chilling because in such an ordinary context, was the case of the study of the progress of cervical cancer at National Women's Hospital in Auckland in the 1960s to1980s. The researchers, although they knew certain conditions were probably pre-cancerous and should have been treated, chose to leave the women patients untreated for years to observe the progress of the disease. Most of the subjects were unaware they were part of any trial. A later judicial inquiry estimated that about 40 women developed invasive cancer and the published study concluded that women in the group receiving no, delayed, or incomplete treatment for cervical cancer were 25 times more likely to develop invasive cancer than women in the control group for whom initial treatment was successful.[17]

These cases, gross deformations of the scientific will to knowledge, show the need to constrain scientific research by outside supervision by ethics committees. But those can only be the last line of defense. Scientific virtue needs to combine and balance commitment to truth and to human welfare, and scientific training needs to treat ethical values as intrinsic to good science, not as some furry add-on to hard science.

And, of course, a scientist must consider the likely uses to be made of the research results that arise. Military research can be morally justifiable, indeed very valuable, if it results in Western civilization maintaining a balance of deterrence over its enemies. The obvious superiority of Western military research was one reason why the Cold War ended without a shot needing to be fired. But the matter has to be thought through in each case, and "not my department" is not an acceptable attitude of a scientist to ethics.

SCIENTIFIC ETHICS: A COMPLAINT

Gather 'round while I sing you of Wernher von Braun
A man whose allegiance

Actually Existing Science

Is ruled by expedience
Call him a Nazi, he won't even frown
"Ha, Nazi, Schmazi" says Wernher von Braun

Don't say that he's hypocritical
Say rather that he's apolitical
"Once the rockets are up, who cares where they come down?
That's not my department" say Wernher von Braun

Some have harsh words for this man of renown
But some think our attitude
Should be one of gratitude
Like the widows and cripples in old London town
Who owe their large pensions to Wernher von Braun

You too may be a big hero
Once you've learned to count backward to zero
"In German, oder English, I know how to count down
Und I'm learning Chinese!" says Wernher von Braun

Tom Lehrer, *Wernher Von Braun* song lyrics (1965).

Science the institution has the capacity to be used either for good or for evil. On balance its effects have been overwhelmingly for good. The increase in life expectancy in the last century has been largely attributable to it. It has given us inventions that have enhanced life in all sorts of ways, from morphine to e-mail. It has cured diseases. It has taught us how to think straight.

There is only one natural human reaction: just be grateful and give science more money.

CHAPTER 13

The Complexity Obstacle to Knowledge
Evolution and Global Warming

S cience, as we have seen, knows a great deal. That is its strongest point. Understanding and admitting its limitations is not one of its strong points.

WHY IS THE SUN HOT?

The sun is hot, as we know, because fusion reactions in it create massive amounts of energy, according to Einstein's equation $E = mc^2$. That answer was discovered in time to be incorporated into the *Encyclopaedia Britannica*'s celebrated 11th edition, of 1911. So what did the 10th edition say? The true answer was "Science has no idea why the sun is hot," but science does not so easily admit its areas of ignorance. Strictly speaking there is no 10th edition of the *Encyclopaedia*. There is a 9th edition (1887), with some "supplementary volumes" added in 1902. Both have an article "Sun." That of 1887 simply avoids the question of why it is hot and confines itself to what is known about its size, mass and structure. The 1902 article makes some calculations of the energy given off and arrives at a large answer, suggesting that if the energy were not replenished, the sun would cool at a rate of 4 percent per annum, which plainly it does not. It mentions the only known possible cause of heating, contraction of the volume of the sun, and calculates that an annual contraction of 6,790 cm would be sufficient to produce the energy. It offers no comment on whether that is possible.

Kinds of Complexity

Some of science's limitations are, indeed, obvious enough. The very small, the very large, and the very old are not easily accessible. As we saw in discussing quantum physics and the Big Bang in chapter 5, it is not easy to determine where in these theories knowledge ends and speculation begins. But in the more medium space and time scales where the great bulk of science lies, the main obstacle to scientific knowledge is complexity.

A typical but comparatively tractable example of how complexity acts to impede knowledge is turbulent flow in fluids, considered in chapter 7. The weather is hard to predict because of the nature of the governing partial differential equations: simple locally, but complex globally because the onset of turbulence quickly pulls apart and "mixes up" the particles.

But it is not as simple as saying "complexity is always bad." Similar examples show also why sometimes complexity is *not* an obstacle to knowledge. In a gas or liquid confined in a container, the particles have very complicated and random paths as they bounce off one another and the walls of the container. That does not prevent us from easily measuring the macroscopic effects of the motion of the particles, such as pressure and temperature. The reason is that those properties are long-term averages of the motions of the particles, and hence the complexity of the motion at the microscopic level is invisible because smeared out. It is a question of scale—if we were content with a long-term average of the weather, we could discover what it was: it is just climate and is relatively stable. But we are small relative to cold fronts and hurricanes, so it is their particularities, not their averages, that we need to know.

There is one other kind of complexity that is relatively amenable to knowledge. When a complex system is modular, there is some hope of understanding and controlling it. Complex human artifacts such as software and machinery are made by assembling modules, just in order for them to be predictable and usable. An automobile has an overall design that is plain to see, then the parts are designed to fulfill a specific purpose in the plan and their insides can be examined on a "need to know" basis by those whose business it is to inquire into technical malfunctions. Software is modular, too: it has a gross structure with parts designed to perform certain straightforward tasks, then within each part the internal structure addresses only that task. Biological and human organizational structures discovered the advantages of modularity too, though they do not implement it as strictly as machinery does (Sales and Accounts can, for example, conspire to subvert a CEO's directives).

Modular, averaged, and even turbulent complexities are tame kinds of complexity. The wild form is found in systems with lots of inaccessible parts of unknown structure and unknown function. Two of the most contentious scientific debates of recent years, the evolution versus intelligent design controversy and the global warming debate, concern such systems. Examining a few aspects of these two complicated debates will show how it is the complexity of the systems involved that is making it hard to achieve the consensus that is normal in other areas of science.

The Theory of Evolution and the Problems of Unintelligent Design

The following survey of evolutionary theory does not pretend to be in any way comprehensive or to establish conclusions. It is an overview with an emphasis on why the particular kind of complexity involved in life and the logical complexity of the theory itself make it hard to establish the theory definitively.

The Darwinian theory consists of four basic propositions:

- Present species have descended over a long period of time from primitive forms (the thesis of "evolution" strictly so called).
- Species change by the accumulation of small changes in individuals, changes that are inherited (with some debates on how small "small" is).
- Those changes happen by chance (that is, as a result of causes unconnected with any aspect of evolution itself).
- Natural selection of the most favorable of these chance small modifications ("survival of the fittest") is the (almost) sole driver of evolution.

In a classic illustration, it is presumed that some giraffes were born at random with genes for longer necks than others; those with the longer necks were able to eat more high foliage and survive better, thus leading to their descendants being more frequently represented in later generations. So the average length of giraffe necks increased over time.

WHAT ARE GIRAFFES' NECKS REALLY FOR?

Abstract: A classic example of extreme morphological adaptation to the environment is the neck of the giraffe (*Giraffa camelopardalis*), a trait that most biologists since Darwin have attributed to competition with other mammalian browsers. However, in searching for present-day evidence for the maintenance of the long neck, we find that during the dry season (when feeding competition should be most intense) giraffe generally feed from low shrubs, not tall trees; females spend over 50% of their time feeding with their necks horizontal; both sexes feed faster and most often with their necks bent; and other sympatric browsers show little foraging height partitioning. Each result suggests that long necks did not evolve specifically for feeding at higher levels. Isometric scaling of neck-to-leg ratios from the Okapi (Okapia johnstoni) indicates that giraffe neck length has increased proportionately more than leg length—an unexpected and physiologically costly method of gaining height. We thus find little critical support for the Darwinian feeding competition idea. We suggest a novel alternative: increased neck length has a sexually selected origin. Males fight for dominance and access to females in a unique way: by clubbing opponents with well-armored heads on long necks. Injury and death during intrasexual combat is not uncommon, and larger-necked males are dominant and gain the greatest access to estrous females. Males' necks and skulls are not only larger and more armored than those of females' (which do not fight), but they also continue growing with age We conclude that sexual selection has been overlooked as a possible explanation for the giraffe's long neck, and on present evidence it provides a better explanation than one of natural selection via feeding competition.

R.E. Simmons and L. Scheepers, "Winning by a Neck: Sexual Selection in the Evolution of Giraffe," *American Naturalist* 148 (1996): 771–786.

The theory is thus logically complex. It is also by its nature rather distant from direct evidence, since large-scale evolution occurs on too long a time-scale to be observable. It is therefore an ideal case study for examining the relation of a complex theory to a large body of circumstantial evidence.

There are some important logical facts to grasp before discussing the evidence for the theory. First, there is some degree of independence between the four theses (for example, Lamarck's theory of the inheritance of acquired characteristics accepted the first and second of these theses but not the third, thus illustrating the independence of the third thesis from the first two). Therefore, one must have evidence for each of the theses separately. Second, the Darwinian theory is not a theory of the origin of life: it concerns only how species mutate into other species, and so it takes inheritance, and thus life, for granted. A credible theory of the origin of life is very hard to find,[1] but that is neither favorable nor unfavorable for the theory of evolution itself. Third, the theory is not, as sometimes claimed, the "foundation for modern biology,"[2] as if biology as a science would come crashing down if it were false. That is so for the obvious reason that the theory of evolution is a theory about *history* whereas anatomy, botany, and so on are about the workings of present-day living organisms. If the theory of evolution were suddenly found false, medicine, for example, would be virtually unaffected. What is perhaps more surprising is that the recent science of genetically modifying organisms has had so little logical connection with evolutionary theory. Although it concerns how to evolve organisms artificially, it seems to have cast very little light on large-scale natural evolution. Finally, it is not very important to the credibility of the core theory of evolution whether or not there is anything in the theories of sociobiology. There have been many justified complaints about the "Darwinian fairytales" dreamed up to "explain" many aspects of human social reality,[3] but an evolutionist could well agree that sociobiology is an embarrassment without needing to give up any core part of evolutionary theory.

If one looks at standard presentations of the "evidence for evolution," such as the *Encyclopaedia Britannica*'s article "evolution" or the National Academy of Science's official book on *Science, Evolution, and Creationism*, one finds that they lead very strongly with such phenomena as the fossil record, geographical distributions of present species, homologies like the pentadactyl limb, and vestigial forms such as gill slits in human embryos. There is reasonable agreement among the descent trees calculated from different kinds of similarities, such as from gross homologies and from blood biochemistry. Those facts are indeed good evidence of evolution, in that they are naturally explained in terms of common ancestry and are not easily explained any other way. The theory makes strong testable predictions in this area: it predicts there will never be a rabbit fossil found in Jurassic strata, and there never has been.

There are indeed a few unsettling aspects to the homology evidence. For example, the pentadactyl limb possessed by the higher vertebrates (sometimes in vestigial form) is especially good evidence of common descent because of the unlikelihood of *five* digits being of any special use and thus explicable in terms of design reasons (whereas four limbs is a natural choice for balance and motion). But that leaves it a mystery as to why the fore and hind limbs have the same number of digits: is it five in both cases because the hind limbs descended from the fore limbs or vice versa? The evolutionary answer must be in some sense "yes": the fore and hind limbs have managed to reuse a common plan. But it is unclear if evolutionary theory has a clear story on how that happened. And attempts to follow limb development and other homologies at the embryological level sometimes lead to awkward questions about the origin of homologies in the adult from different parts and processes in the embryo.[4]

Nevertheless, by and large the standard evidence for evolution is very strong. But it is important to notice that that evidence bears only on the *first* of the four theses of Darwinism, that of common descent from primitive forms. It has no bearing on the other three theses, which concern the *mechanism* of evolution. So those theses need to be established by their own bodies of evidence. That evidence is, in brief:

- The evidence that evolutionary changes are generally small comes from genetic experiments and artificial breeding. However Darwin's thesis that evolution is always continuous has been modified as genetics has understood the role of the discrete DNA structure in inheritance. It is known that there must have been some large "jumps" of certain kinds in evolution—for example, the tripling of genetic material that has caused cultivated wheat to have three times the number of chromosomes as its wild ancestor, or the incorporation of mitochondria in the cell.

Unfortunately, there is a good reason why Darwin preferred continuity—the more one invokes jumps in "punctuated equilibrium" theory to, say, explain gaps in the fossil record, the less theory one has left, since a jump to a functioning new form is an improbable and unexplained coincidence—and the bigger the gap, the more improbable the jump.

- The evidence that genetic changes are due to chance is that variation among offspring is observed, but no substantial connections have been found between the changes and what would be useful for the organisms in their environment. Despite intensive search, no important case of inheritance of acquired characteristics has been found. The known rates of some mutations such as that for hemophilia in humans, and the known ability of radiation to increase mutation rates, also suggest the randomness of genetic changes. It is true that recent research on the ways that activation of DNA can be turned on and off has revealed some possibilities of "epigenetic" changes in gene expression that can persist for some generations without being incorporated into the DNA, but it seems unlikely this will make much change to the big picture of evolution.[5]

- The logically weakest point in evolutionary theory—and this must be so in its nature—is the last thesis, that the mechanism of inherited random variations with natural selection, which can undoubtedly cause *some* evolution, is sufficient to explain *all* of it. Probably nearly everyone, on first reading about Darwin's theory, has an initial feeling that not only is it such a good idea that it deserves to be right (as Thomas Huxley put it, "How extremely stupid not to have thought of that!"[6]) but that it *must* be right—given variation in offspring, surely natural selection *must* act to cause evolution? That is a dangerous illusion of certainty. Although it shows there must be evolution, it says nothing about how much there will be. It may be, for example, that it is easy to breed a bigger potato by selecting from the existing variability in the potato gene pool, but that transforming a potato into a yam requires a fundamentally different process. Darwin was well aware of the problem, which is why the first chapter of the *Origin of Species* is full of mind-numbing detail on the breeding of pigeons. For the main argument for the sufficiency of the chance-and-selection mechanism to explain large-scale evolution over a long time is an extrapolation from its ability to explain the artificial evolution of varieties over a short time— and our inability to think of any other serious theory. As arguments go, extrapolation far beyond the data and an inability to think of alternatives are not strong.

That covers the main lines of the evidence for the Darwinian theory of evolution, as amended. It is substantial, but, as always, "Hear the other side." We must consider the objections.

Darwinism's record on dealing with objections is not impressive. Objections have tended to jump from the status of being cavils put forward by ignorant religious polemicists to being serious, but now solved, objections, without passing through an intermediate phase of being serious but unsolved objections. A cartoon history goes something like this:

- In Darwin's time, Kelvin argued that the earth was still hot inside so it could not be very old, certainly not old enough for the "eons of geological time" needed for evolution. The evolutionists suggested not worrying about this argument and hoped that something would turn up. Something did turn up—a totally new form of energy capable of generating heat inside the earth. But that showed that the objection must have been a good one at the time.
- Darwin had a "blending" or continuous theory of inheritance that implied that a beneficial mutation would tend to

223

be reabsorbed into the mass of the unmutated population before it had a chance to be selected for. Mendel's solution in terms of a discrete gene theory of inheritance showed the objection was a good one in the first place.

- The obvious problem of gaps in the fossil record was attributed originally to the lack of effort in looking for them; Darwin had a stroke of luck when *Archaeopteryx* was found only two years after the publication of *Origin of Species*. The rate of discovery of "missing links" in the time since, though not exactly as much as expected, has been sufficient to maintain optimism.

- The early Darwinists' theory of competition as "nature red in tooth and claw," or at least of a ruthless competition for resources, left it impossible to explain how altruism could have evolved. The development of group selection, game theoretic, and "selfish gene" theories in the mid-twentieth century allowed this problem too to move from the swept-under-the-carpet box to the triumphantly solved box.

The problems that remain serious matters of contention concern the ability of evolution to explain the observed complexity of organisms, given the time available for evolution. The nature of the case makes it extraordinarily hard for either side in the debate to establish its position conclusively.

"Given sufficient time," Richard Dawkins says, "the non-random survival of hereditary entities (which occasionally miscopy) will generate complexity, diversity, beauty, and an illusion of design so persuasive that it is almost impossible to distinguish from deliberate intelligent design." That is true, but is the time actually given anything like sufficient? Dawkins says, "not a problem either—except for human minds struggling to take on board the terrifying magnitude of geological time."[7] But geological time is not terrifying. The age of life on earth is only about four billion years. That is not a very long time for a random search process such as Darwinian evolution to achieve results. A fully random process such as monkeys typing will not make any progress with *Hamlet* on that sort of time scale. Of course, a Darwinian process that preserves, advances, and builds on them can do much better, but how much better needs to be established mathematically. Ideally, one would like to have an estimate of how much complexity Darwinian evolution could accumulate (in, say, bits per generation), and an estimate of the complexity of an organism such as a human. Then the two could be compared. Although there have been some attempts in that direction, such as Kimura's estimate that in the last 500 million years complexity has been increasing at 0.29 bits per generation,[8] the nature of biological complexity makes it impossible to measure the relevant kind of complexity, at least with present conceptual technology.

One might be tempted to regard the DNA as a kind of blueprint or code of the organism, and hence to measure the complexity of an organism by counting the number of base pairs in its genome. (The base pairs are the "rungs" on the DNA strands that carry the "information"—a change in one of them is the simplest kind of mutation; the human genome has about 3 billion of them, or nearly one for every year of evolution.) That is a bad idea for a great number of reasons. In one way it is an overestimate, as a large proportion of the DNA is "junk," and a change in it makes no difference to the organism. In another way it is a gross underestimate of the complexity of organisms, since there is much more in the organism than is specifically coded in DNA: for example, the human brain has about 10 billion neurons, each with an average of 7,000 connections to other neurons, so obviously DNA, at most, codes for the growth of neurons rather than for all the completed neurons and connections. And in any case, the relation between genetic material and expressed characteristics is far from one-to-one. But those phenomena are only symptoms of a much more fundamental problem. The relation of DNA to organism is not at all like a code or blueprint. The DNA does "issue instructions," in a sense, but they are instructions for the action of proteins. (Even that is too simple, since the proteins act in many cases through their shape when folded, and protein folding is not itself encoded in DNA.) The organism is the result of the interactions of those actions, with one another and with the complex environment that must be provided for the DNA to act, namely the cell (and what lies outside the cell). So the relation between DNA and completed organism is more like the relation of the rules of chess to chess strategy than the relation of blueprint to machine. The small number of rules in chess generates nearly infinite possibilities in strategy (while another game with the same number of rules might be much simpler). Even if, by and large, more DNA allows for more possibilities of complexity in an organism, the amount of DNA is entirely inadequate as a measure of an organism's complexity. The question, "Could something as complex as humans have evolved by chance and natural selection in four billion years?" remains a good one, but there is no immediate prospect of answering it one way or the other.

In view of that impasse, the intelligent design opponents of Darwinism tried another tack, attempting to exhibit "irreducibly complex" biological organs that, because of the interrelation of their parts, could not possibly (or only with extreme improbability) have been generated by the Darwinian process. The idea of the problem is an old one. Darwin said, "If it could be demonstrated that any complex organ existed which could not possibly have been formed by numerous, successive, slight modifications, my theory would absolutely break down. But I can find out no such case."[9] He then spent a good deal of effort explaining, with some success, how the eye could have evolved gradually from a primitive light-sensitive spot. There was considerable subsequent argument on the "problem of the incipient stages of useful structures," debating whether, for example, half a wing was worse than no wing at all.

Recent debate has focused on a few clear examples. One of them, advanced by the prominent ID advocate Michael Behe, is the case of the bacterial flagellum. Evolution has never managed to invent the wheel, but it has achieved the propeller. On anyone's view, it is a remarkable piece of machinery, with some claim to be the "most efficient machine in the universe."

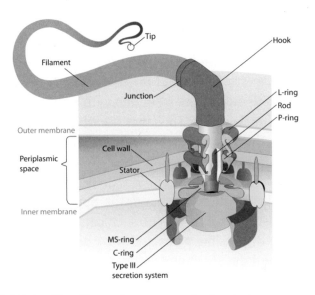

FIGURE 13.1   Flagellum of gram-negative bacteria.

A chemical pump rotates the flagellum at variable speeds up to 1,000 revs per minute and propels the bacterium forward at up to 60 cell lengths per second. Rotary motors require complex engineering and many interlocking parts. A complex process of assembly is also needed. Behe argues that the complexity is "irreducible" in the sense that if the mechanism lacks any major part, it cannot function. Therefore, he says, it cannot possibly have evolved bit by bit as Darwinism asserts.[10] He writes:

> *An irreducibly complex system cannot be produced directly by numerous, successive, slight modifications of a precursor system, because any precursor to an irreducibly complex system that is missing a part is by definition nonfunctional.... Since natural selection can only choose systems that are already working, then if*

*a biological system cannot be produced gradually it would have to
arise as an integrated unit, in one fell swoop, for natural selection
to have anything to act on.*[11]

It needs to be understood what a difficult task it is, logically, for either side in the debate to establish its position on
this question conclusively. The evolutionist side would need
to demonstrate a possible smooth path (or one smooth but for
occasional gaps, which would themselves each need explanation) of organisms passing from one with no flagellum to one
with a working flagellum, such that at each point on the evolutionary path the organism was viable *and* such that each point
was better adapted than the point before (so that natural selection could drive evolution forward along the path). That is a
very big task. On the other hand, the opponent of evolution
must demonstrate that there is no such path. Given the vastness
of the space of possible paths and the poor understanding of
that space (and the possibility of jumps, of debatable probabilities) that is also a nearly impossible task.

The official evolutionist answer to Behe makes much play with
the notion of "exaptation." An exaptation is a reuse or coopta-
tion of a machine or machine part for some purpose other than
its original one. Kenneth Miller, a leading figure in replies to
Intelligent Design arguments, suggests that close examination
of the flagellum shows that some of it is a modification of an
organ that did have another purpose, namely a "Type III secre-
tory system," whose purpose is for bacteria to inject toxins into
their host cells. That shows, Miller says, that the complexity of
the flagellum system is not "irreducible" as Behe claimed but
reducible, that is, that some part of it was already useful.[12] A
recent review of the field points to the variety of different flagellar mechanisms and to homologies between several parts of
the flagellar system and other systems.[13]

Exaptation is a good idea, certainly. In particular, it provides
the possibility of a jump in evolution that is of reasonably high

probability—since the incorporation of a working part in a new mechanism is a large discrete event but not very surprising. But it only makes a moderate impact on the problem of finding an evolutionary path to a complex mechanism. An exaptation event is a one-off, and most of the path, up to and after the event, still has to be found. None of the responses to Behe seem to be close to offering that.

And as with other answers to objections to Darwinism, if that is the answer, then the question was a good one. If the "irreducible complexity" objection does need exaptations to be inserted into the continuous evolutionary paths supposed by the original version of Darwinism, then it was a good objection before the theory was so modified.

Perhaps the most scientifically useful outcome of the ID debate has been to focus attention on how little we do understand "artifactual" complexity, the kind in machines and biological flagella in which different working parts cooperate to produce an outcome. Miller has been known to wear a miniature mousetrap as a tie pin, by way of showing how one machine can be "exapted" to another purpose. But if we did understand the complexity of machines, we would know how to use evolutionarily inspired search algorithms to design a better mousetrap. There is currently no known way of designing machines using random search, because we do not understand the complexity of machines well enough to specify the search space.

Global Warming: Multiple Causes and Variable Timescales

As in the case of evolution, global warming theory consists of several independent theses, each in need of its own sources of evidence. Its main propositions are:

- The world has been warming unusually in recent decades.
- There are no known natural causes of this warming.

- The burning of fossil fuels is necessary and sufficient to explain the warming.

(There are further debates about the effects and costs of proposed actions, but we will not attempt to consider those.)

Again, like evolution, the first thesis is considerably easier to establish than the later ones, which concern causes. The recent rise in global mean temperature is apparent in the graph supplied by the IPCC (Intergovernmental Panel on Climate Change, the official body reporting on climate change):

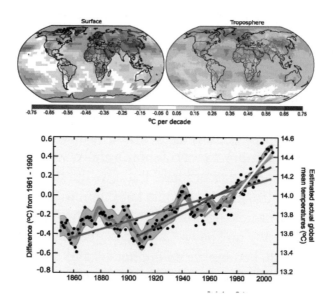

FIGURE 13.2   Climate Change 2007: The Physical Science Basis. Working Group I Contribution to the Fourth Assessment Report of the Intergovernmental Panel on Climate Change. Figure TS.6. Cambridge University Press.

This graph summarizes and depends on the reliability of a great deal of work on data and considerations about possible biases in it (for example, whether the location of temperature recording stations near cities has created a "heat island" effect). It seems the data is solid and there really is a notable rise in temperature in recent decades. There are various regional variations that need to be understood, such as increases in the poleward transport of tropical heat.[14]

The drop in the period 1940–1970 needs explaining, if the rise is to be attributed to an increase in carbon dioxide since the Industrial Revolution. The consensus of opinion is that the dip is probably due to aerosols (particles such as soot, etc.) from dirty coal-burning and industrial processes reflecting more sunlight. That is plausible but the effects of different kinds of aerosols are very variable.[15] A letter in *Nature* of May 2008 explains a sudden drop in 1945 as a result of measurement error, caused by a change in the way of measuring sea surface temperature,[16] a suggestion that does not inspire confidence in the record (even though the general coldness of the late 1940s is well-established).

Paleoclimate is important to understand, since if we cannot explain past patterns we cannot be confident we understand present patterns, and because what is unusual or unnatural depends on the natural variability in normal conditions. If we go back a couple of thousand years, there are some slightly warmer and cooler periods, but nothing as large as the recent increase. There has been much debate about the "late medieval warm period" in Europe, but the general picture for the whole globe is not so clear.[17] Further back, there are bigger ups and downs, especially ice ages (the last one only 11,000 years ago). It is believed that they were triggered by changes in solar radiation, which act on much longer time scales than the recent warming. The complex paleoclimatic record and the poor understanding of the causes of its fluctuations form one of the main areas in which climate science suffers from the curse of complexity.[18]

The next question concerns causality, which is, as usual, harder. Is the temperature increase caused by the burning of fossil fuels since the Industrial Revolution?

One can see in the records a clear rise in atmospheric carbon dioxide and other "heat-trapping" gases such as methane since 1800 and especially since 1950. Physics then shows that they act as "greenhouse gases," that is, they trap in the atmosphere heat from the earth that would otherwise escape into space. That in principle should be a simple matter of the physics of the globe, needing, for example, no aggregation of potentially dubious computer models of regional climate. Yet it is not as simple as that since the direct rise in temperature attributable to those gases is very much less than that observed. The main part of the increase is attributed to an increase in water vapor in the atmosphere, itself caused by the greenhouse warming, which multiplies the warming effect several times. How many times, exactly, is not easy to say, since at this point climate science runs into its second major obstacle arising from complexity. Cloud formation is very difficult to model, and its interactions with aerosols, global heat transport, the trapping of heat, and the reflection of solar radiation create large uncertainties as to the total effect of increased water vapor.[19] It is also in the nature of a feedback effect (warming causes more water vapor which, in turn, causes more warming) that it is hard to predict how far it extends. It is also true that any other cause of warming would produce the same kind of feedback effect (though there are no plausible candidates). The argument that the degree of warming to be expected is approximately what is observed is reasonable, but the quantitative uncertainties are substantial.

To understand the effect of burning fuels on the amount of carbon dioxide in the atmosphere, it is necessary to gain an approximate quantitative overview of the carbon cycle:

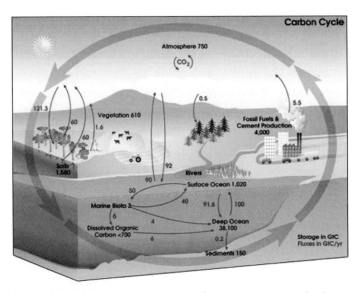

FIGURE 13.3   NASA Earth Observatory's Carbon Cycle diagram.

It can be seen that the amount of CO_2 being released into the atmosphere by burning fuels is quite large compared to the amount in the atmosphere. The atmosphere does not contain a great deal of matter, so any large influx is dangerous. On the other hand, the amount of CO_2 from fuels is not very large compared to the natural circulation of carbon into and out of vegetation, soils and the oceans. That means that uncertainties over historical changes in those numbers can impact severely on the accuracy of estimates of the effect of fossil fuels. The rotting of vegetation in soils is particularly hard to estimate. Many forests have been cleared since the Industrial Revolution and global warming itself can cause faster rotting and release of CO_2,[20] while the effects of the rapidly warming Arctic tundra are also hard to estimate.[21] Previous IPCC reports attributed a considerable fraction of the human-caused increase in CO_2 to such processes but the latest report is much less committed.

Again it would be good to understand paleoclimate to see if we properly grasp the relation between temperature and CO_2. Data is available from one of the great observational projects of recent science, the 3km-long Vostok ice core, taken from inland Antarctica in 1998. From the air bubbles trapped in the ice, it is possible to reconstruct temperature and CO_2 levels (in Antarctica) for the last 400,000 years. The results are:

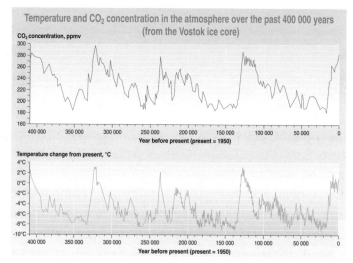

FIGURE 13.4 © Philippe Rekacewicz, UNEP/GRID-Arendal.

The covariation of temperature and CO_2 is clear. During ice ages, there is much less CO_2 in the atmosphere than there is now (though nowhere is CO_2 as high as at present). This diagram is sometimes presented as evidence that CO_2 does cause rises in temperature, for example, in Al Gore's documentary, *An Inconvenient Truth*. But that is not correct. A close inspection of the data shows that the warming comes first and the CO_2 rise later, with a lag of about 800 years, and that the peaks and troughs also approximately match those of incoming solar energy.[22] The IPCC report's section on Paleoclimate admits that the reasons for this are unknown, though quite likely connected with Southern Ocean overturning.[23] Although this mechanism is probably different from what is happening in the current peak, the situation is not very satisfactory.

Finally, is there global warming on Mars? Comparison with a parallel case should be instructive, since there are no Martians to burn fossil fuels. There is global warming on Mars.[24] However, most experts think it is due to huge dust storms that change the planet's albedo.[25] The cause of the storms is not very clear, nor is whether there is any long-term trend in them. So Mars is probably not a comparable case after all. It is all very complex....

Readers may follow the latest on these issues on the Climate Debate Daily website.[26]

In both evolution and climate change, the majority view of the scientific experts is well ahead. In neither case is there any known coherent alternative. But the complexities of the evidence are such that a higher standard of politeness to skeptics who raise serious problems would be well-advised.

CHAPTER 14

Is That All There Is?

The argument "Science has explained or will explain everything there is to explain" is potentially a good one. But it
is not a scientific one—science itself does not make any findings about its limits. The argument is a philosophical one: of
the various phenomena we know, either science has explained
them or there is good reason to believe future science will prove
capable of explaining them (at least in principle). It would follow
that understanding the universe and our place in it would not
require any non-scientific entities or principles such as gods,
disembodied minds, distinctively ethical principles, or similar.

What is not such a good argument is one often given in support of the "science is all there is" conclusion, which arises from
the past successes of science. It is sometimes said that the fact
science has shown we do not need a Corn God to push up the
crops, angels to move the planets, or special vitalist principles to
explain life suggests that scientific progress has a kind of unstop-
pable momentum that will eventually sweep over all remaining
pockets of resistance. Besides sharing the weakness of all extrapolation arguments, that way of putting it invites also considering
the counterargument: now that science is so advanced and we

understand its methods so well, any phenomena that remain outside its scope probably do so for some good reason of principle.

There are two areas in particular where there is reason to think that the phenomena are not only beyond the present reach of science, but are in their nature inexplicable by science. They are consciousness and ethics.

Consciousness and Qualia

What is it like to be a zombie? A zombie is, by definition, a being physically identical to a human but without any conscious experience. Maybe the laws of nature prevent there being zombies, in that anything physically the same as humans in fact gives rise to consciousness. Zombies are a philosophers' thought experiment, and all that is necessary is that it is conceptually possible that there should be such things.[1]

It is not like anything to be a zombie, since a zombie has no conscious experience. It is intelligible to ask, "What is it like to be a bat?" (though hard to answer): "an organism has conscious mental states if and only if there is something that it is to *be* that organism—something it is like *for* the organism."[2]

The point of the thought experiment is to make as precise as possible our intuition that there is something entirely unique about consciousness, something that suggests it is not the kind of thing that could possibly be physical. As one philosopher puts it, the human brain is "just the wrong kind of thing to give birth to consciousness. You might as well assert that numbers emerge from biscuits or ethics from rhubarb."[3] Other thought experiments that produce the same conclusion are "What Mary knew," concerning the possibility of having all physical knowledge about a situation but still lacking the "raw feel" or *qualia* of colors, and the inverted spectrum, which asks whether one person's spectrum of perceived colors might in fact be the inverse of another's (which could of course not be revealed by external questioning). Each of them is intended to dramatize and reinforce our strong feeling that the realm of consciousness and the realm of physical reality are of entirely different natures.

WHAT MARY KNEW

Mary is a brilliant scientist who is, for whatever reason, forced to investigate the world from a black and white room *via* a black and white television monitor. She specialises in the neurophysiology of vision and acquires, let us suppose, all the physical information there is to obtain about what goes on when we see ripe tomatoes, or the sky, and use terms like "red," "blue" and so on. She discovers, for example, just which wave-length combinations from the sky stimulate the retina.... What will happen when Mary is released from the black and white room or is given a colour television monitor? Will she *learn* anything or not? It seems just obvious that she will learn something about the world and our visual experience of it. But then it is inescapable that her previous experience was incomplete. But she had *all* the physical information. *Ergo* there is more to have than that, and Physicalism is false."

Frank Jackson, "Epiphenomenal Qualia," *Philosophical Quarterly* 32 (1982): 127–36; novelized in David Lodge, *Thinks...* (New York: Viking Adult, 2001), p. 53 and ch. 16.

The answer of materialists or physicalists—those who think the mind is nothing over and above the brain—must be that this intuition, strong or not, is an illusion. D. M. Armstrong, author of the classic *A Materialist Theory of the Mind*, suggests comparing it to the "headless woman illusion." A magician places a woman against a dark background with a dark cloth over her head. It appears to the audience that she has no head. That is, the audience's perceptual systems are set up to infer from "I do not perceive the woman has a head" to "I perceive the woman does not have a head." Similarly, Armstrong suggests, we illegitimately pass from the true "I am not introspectively aware that sensations are brain processes" to the false "I am introspectively aware that sensations are not brain processes."[4]

That is an ingenious answer, but it is not clear that it entirely solves the problem. The point of Descartes's "I think therefore I am" is that an illusion about one's own existence is impossible, because something must be having the illusion. An awareness of one's own sensation, as sensation, is basic. And not only basic to philosophy, but to science, which depends entirely for its ultimate inputs on observations, that is, human sensations. It would not be easy, logically, for science to start with sensations, infer the existence of a physical world, and then cut off the branch it was sitting on by deleting the sensations.

The Science of Consciousness

That is not to say that there could not be, in some sense, a science of consciousness. Consciousness may be hidden from view (of others) but there are repeatable phenomena in it and in the relation of consciousness to physical phenomena that are susceptible to scientific study.[5]

The extinction of consciousness on physical death, and temporarily in response to severe blows to the head or excessive ingestion of certain substances, are reliable correlations between consciousness and the physical realm that have been known since time immemorial, and there has been extensive work on the connections between damage to brains and the effects on consciousness.[6] One of the most solid contributions of the nineteenth century to scientific psychology was the Weber-Fechner law, which says that (by and large, and to a reasonable approximation) the perceived change in a quantity is a logarithmic function of the quantity itself. For example, Weber tested the least perceivable change in weight held by a blindfolded man and found it was proportional to the *relative* increase in weight.

For any weight (in the range where weights can easily be felt), an increase of 10 percent is felt to be much the same increase. Similar results hold for brightness of light and loudness of sound. What is being compared is of course *reported* change in perception, but as these are reasonably constant across observers, it seems reasonable to suppose that the reports are a generally accurate account of what is in the consciousness of the subjects, and hence that the Weber-Fechner law and its descendants are a genuine instance of a scientifically established correlation between consciousness and physical causes.

In recent years, the classic results of this kind are those of Benjamin Libet on the timing of voluntary actions, which have been taken to be evidence against the existence of free will. A subject watches a dot that travels steadily around a circle and is used as a timer. The subject is also wired with electrodes on the scalp to record brain activity. He or she is then asked to perform, at a time of his or her choice, some simple voluntary action such as pressing a button—and to note the position of the dot on the timer "when he or she was first aware of the wish or urge to act." The timing of the voluntary decision could thus be compared with the time of the first brain events associated with putting the decision into action.

It is found that the first brain events to initiate the action occur *before* the conscious decision—about 0.3 seconds before, a long period as brain events go. A natural conclusion to draw is that the conscious decision to act cannot be the cause of the action, since the action has already been put in motion by the brain some time before the conscious act. Thus our belief that we consciously make free decisions is an illusion.[7]

That interpretation of the results is controversial. Libet himself believed that although the decision was initiated involuntarily, the conscious will maintained a power of veto and could later interrupt the action or allow it to proceed. There is also room for doubts about whether noting the position of the timer really does correctly time the voluntary act, or whether some period should be allowed for an act of deciding to register the position of the timer.

Nevertheless, the experiments show the genuine possibility of scientific work on the connections of consciousness to physical reality—all that is needed is that consciousness and physics be in the same time, and that there be credible and reproducible reports by subjects on what is in their consciousness.

None of that serious work on consciousness and physics has any tendency to support materialist views of the mind.

PHENOMENOLOGY: CONSCIOUSNESS DESCRIBED "FROM THE INSIDE"

But it is not necessary that they [humans, dogs etc] and other objects likewise should be present in my *field of perception*. For me real objects are there, definite, more or less familiar, agreeing with what is actually perceived without being themselves perceived or even intuitively present. I can let my attention wander from the writing table I have just seen and observed, through the unseen portions of the room behind my back to the verandah, into the garden, to the children in the summer-house, and so forth, to all the objects concerning which I precisely "know" that they are there and yonder in my co-perceived surroundings...

Edmund Husserl, *Ideas: General Introduction to Pure Phenomenology*, trans. W. R. Boyce Gibson (London: George Allen & Unwin Ltd., 1931), 137.

Ethics

In a famous passage in his *Treatise of Human Nature*, David Hume denies the derivability of "ought" from "is." Matters of fact, he says—the kind of facts discovered by science—cannot by themselves determine questions of ethics. The distinction is a good one. What is—in the sense of science—does not determine what ought to be.

HUME ON THE IS-OUGHT GAP

I cannot forbear adding to these reasonings an observation, which may, perhaps, be found of some importance. In every system of morality, which I have hitherto met with, I have always remark'd, that the author proceeds for some time in the ordinary ways of reasoning, and establishes the being of a God, or makes observations concerning human affairs; when of a sudden I am surpriz'd to find, that instead of the usual copulations of propositions, is, and is not, I meet with no proposition that is not connected with an ought, or an ought not. This change is imperceptible; but is however, of the last consequence. For as this ought, or ought not, expresses some new relation or affirmation, 'tis necessary that it shou'd be observ'd and explain'd; and at the same time that a reason should be given; for what seems altogether inconceivable, how this new relation can be a deduction from others, which are entirely different from it. But as authors do not commonly use this precaution, I shall presume to recommend it to the readers; and am persuaded, that this small attention wou'd subvert all the vulgar systems of morality, and let us see, that the distinction of vice and virtue is not founded merely on the relations of objects, nor is perceiv'd by reason.

David Hume, *A Treatise of Human Nature,* Book III, section i.

Some of Hume's more enthusiastically atheist followers have taken this to mean that when it comes down to it, there is no such thing as ethics. John Mackie held an "error theory" of ethics, that our feelings of right and wrong are illegitimate projections onto the world of our habits as determined by our evolutionary history and individual training.[8] Many writers on evolutionary ethics have not quite gone to Mackie's extreme of entirely denying that there is really a right and wrong, but why they stop short is often not clear from their reasoning.

That conclusion, whether in extreme or moderate form, does not follow from what Hume says. He merely points out that factual matters—and that includes facts about what is conducive to survival, what the customs of our tribe are, and what God has commanded—do not have ethical consequences by themselves. So the source of ethics (if there is such a thing as ethics) will have to be sought elsewhere.

Before inquiring what the source of ethics might be, let us consider how Hume's lesson has not been learned by some of the scientifically-inclined writers of recent times, who have hoped to find "lessons" in science to fill the moral void left by the supposed departure of religiously based ethics.

One thing that Hume's distinction rules out is flabby moralizing on the humility we should feel at the scientific discovery that we live in a small solar system in an obscure corner of the universe. Stephen Hawking repeats a common theme in writing, "We are such insignificant creatures on a minor planet of a very average star in the outer suburbs of one of a hundred thousand million galaxies. So it is difficult to believe in a God who would care about us."[9] Size and position are not morally relevant properties, for the same reason as the skull measurements of Nazi "racial science" were not relevant to the worth or rights of the people on whom the measurements were made. Our size relative to that of the universe says nothing about our worth, and the "insignificance" of life on earth is not a fact science knows or could possibly know. Hume's is-ought distinction rules out any such crossing of the boundary from fact to value.

Almost equally dubious are claims that science gives us quite enough to "wonder" about without our looking to religion. Richard Dawkins begins *The God Delusion* by approving of a "quasi-mystical response to nature and the universe" that is common among scientists, and aspires to "touch the nerve-endings of transcendent wonder that religion monopolized in past centuries."[10] Wonder, in the sense of curiosity and surprise, is a natural human reaction to the remarkable facts science reveals, but it is a reaction *of ours*, not a quality of the material universe. Science reveals no more than that the universe is a pretty ordinary pile of rocks with a lot of space in between. Galaxies or spiders are not in themselves "wonderful," or "remarkable" (just possibly "improbable," but that lacks the emotional quality of "wonderful").

Hume's is-ought gap just as firmly rules out any kind of "evolutionary ethics," in the sense of an ethics based on, or reducible to, facts about evolution. If our evolutionary history has equipped us with any habits, that in itself is no reason at all why we should approve those habits or follow them. Perhaps evolution has helped give us habits that we regard as moral, such as altruism or survival instincts. Some evolutionary theorists felt a rosy glow when it turned out that nature need not be as "red in tooth and claw" as the early Darwinists supposed, but that altruism too could be shown to have evolutionary advantages. But any attempt to actually base ethics on its evolutionary history faces a problem when evolution comes up with less savory characteristics. Rape and ethnic cleansing offer obvious advantages for "selfish genes" that wish to spread themselves to future generations—advantages well realized by Genghis Khan, who has an estimated 16 million living descendants.[11] But rape and genocide are evils and their role in evolution is entirely incapable of providing an excuse for them. Hume is correct—there is no possible inference that "Whatever is, is right." Whatever evolutionary traits, tribal customs or individual acts exist, moral criticism of them from an outside standpoint is possible and necessary.

There is likewise no mileage to be gained from arguments about ethics arising from the fact that ethical judgment is performed by real brains. A version of biologically inspired moral skepticism particularly untroubled by philosophical sophistication appears at the beginning of E. O. Wilson's very influential *Sociobiology*:

> ... *self-knowledge is constrained and shaped by the emotional and control centers in the hypothalamus and limbic system of the brain. These centers flood our consciousness with all the emotions—hate, love, guilt, fear, and others—that are consulted by ethical philosophers who wish to intuit the standards of good and evil. What, we are then compelled to ask, made the hypothalamus and limbic system? They evolved by natural selection. That simple biological statement must be pursued to explain ethics and ethical philosophers, if not epistemology and epistemologists, at all depths.*[12]

The argument is: "We cannot know ethical truths (if there are any) except through the urgings of our back-of-brain plumbing, therefore, we cannot know ethical truths at all." This is of the same form as "We cannot know mathematical truths except through the calculations of our frontal cortex, therefore, we cannot know mathematical truths at all."

Its invalidity is obvious. It is an example of Stove's "worst argument in the world," discussed in chapter 3.[13]

The Science of Ethical Behavior and Intuitions

With the philosophy out of the way, there is no problem with noticing that science has discovered interesting facts about ethical behavior, as actually implemented in real people and real societies. "How is it that altruism survives and selfishness often has bad consequences?" is a reasonable question, answerable by scientific investigations about how interpersonal interactions work.

Game theory provides a model for how ethical behavior is often self-sustaining (but sometimes is not). In the classic scenario of the Prisoners' Dilemma, two prisoners, in fact guilty of collaborating in a crime, are interrogated separately. The interrogator makes each an offer: parole if you confess and the other does not; 1 year's jail if neither confesses; 20 years' jail if both confess; life if you do not confess and the other does.

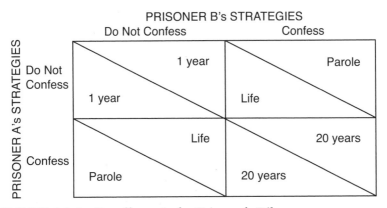

FIGURE 14.1   Payoff matrix for Prisoner's Dilemma.

The ordering of severity of the payoffs is designed to create a conflict between the self-interest of each prisoner and what would be better for the pair of them: each is under pressure to "save his own skin" by confessing, but knows that the other is under the same pressure, and that if they both confess, they will be worse off than if they both refuse to confess. The best strategy in the single game is normally thought to be to confess, on the grounds that that has better payoff whatever the other prisoner does. But most of the interest in the topic revolves around iterated prisoners' dilemma, where a similar game is played many times and each prisoner can observe the other's past behavior. The best strategy is then "tit-for-tat": cooperate (with the other prisoner, that is, do not confess) in the first round, then do as the other did in the previous round. This strategy gains the benefits of cooperation without exposing the player to the costs of gullibility. The optimality of the "tit-for-tat" strategy remains across a wide range of variations on the game.

The original applications of the game were to scientific questions, such as analyzing cooperative behavior in business and showing how altruism was compatible with the Darwinian theory of evolution. But it was not long before popularizers of sociobiology and some philosophers began to draw ethical conclusions. The philosophical significance was normally taken to be in favor of ethical egoism: altruism is explained away as "really" self-interested action, on the part of either the individual or his or her "selfish genes."[14] That cannot be right, for the reasons Hume gives: it would replace "ought" with "is." But it does show why unethical behavior can often be punished, as a society of cooperative people ostracize and isolate someone who takes the selfish option. (But unfortunately, there is one variation that makes cooperation no longer optimal: in "iterated prisoners dilemma with finite horizon"—in which it is known when the game ends—the last few moves reduce to the single-game case, and it can be optimal to turn very nasty at the last moment.)

Science has also come up with some disturbing results on the effects of circumstances on real ethical behavior. Work such as the Milgram experiments and the Stanford Prison Experiment have made reproducible findings on the role of obedience to authority in committing unethical actions. In the original 1961 Yale experiment by Stanley Milgram, participants believed they were administering electric shocks to another subject (whom they could not see but could hear apparently screaming). On the orders of the experimenter, a stern figure in a white coat standing beside them, they increased the (apparent) shocks to the subject. Two-thirds of subjects were prepared to increase the shocks to a level they believed could be lethal to the subject.[15] In Philip Zimbardo's 1971 Stanford prison experiment, volunteers were divided into "guards" and "prisoners" in a mock prison. Both sides adapted quickly to their roles, the "guards" becoming so sadistic and some of the prisoners so trauma-tized that the experiment was stopped six days into its planned 14-day duration.[16] Those results are alarming, and undoubtedly show something about the darker side of human nature and the effects of social pressures on individual actions.

Science has also made progress on the question of the uni-versality of moral intuitions. Marc Hauser's *Moral Minds* argues for a kind of "universal moral grammar"—in the mind but barely accessible to consciousness, like the universal principles of grammar common to all languages—on the basis of surveyed responses to moral dilemmas like this:

Case 1: A runaway trolley on a track is about to hit and kill five people who are on the track if it continues. But you, a bystander walking nearby, come upon a switch that you can flip to turn the trolley onto a side track where it will kill one person. You are asked, is it permissible to flip the switch?

Case 2: Five people have been rushed into the emergency department of a hospital; each one needs a transplanted organ to live. A completely healthy man walks in. Taking his organs will save the five but he will not survive. Is it permissible for a surgeon to kill the healthy man for his organs?

A simple "Though shalt not kill" rule would say both cases are impermissible. A simple utilitarian rule, ruling that the greatest benefit for the greatest number should determine what is right, would make both cases permissible. But research shows that the great majority of humans find the first case permissible and the second not. That is true whatever the religious or non-religious upbringing of the respondents, and largely true across cultures (substituting where necessary culturally appropriate examples using crocodiles). Hauser concludes that humans have a hard-wired "moral sense" that develops in childhood in essentially only one possible direction, and involves considerations of intention, distinctions between actively doing and permitting, and the like.[17]

The interpretations of this recent research remain controversial. But there are two things that can be said about this and any similar research on ethical behavior, "neuroethics," cross-cultural commonalities or differences in ethics, and any other scientific research on ethics. First, they are genuine science and can give us valuable insights into how and why we behave ethically (or not). Second, to the extent that they are science, they are not *ethics*. Hume's distinction means that their descriptions of what is the case can neither support nor undermine any considerations of what ought to be.

Real Ethics

So if science cannot give us our ethics, where should we begin to look?

When we are confronted with pictures of genocide victims dug up (e.g. of Srebrenica) we know that "those were people like us, and something terrible happened to them." Our emotional reaction gives us an immediate perception of the violation and destruction of something of immense value, a human life. It is gross violations of the right to life that most immediately force upon us a sense of the objective inviolability of human worth, which is the foundation of our knowledge of the objectivity of ethics. To be skeptical about something as ethically basic as the terribleness of evil suffered by the victims of genocide would be not only wrong but itself an evil act against the victims of evil.[18]

We need our emotions, not just a purely scientific rationality, to know the terribleness of evil. Anyone who does not have an immediate revulsion to images of genocide is autistic. He suffers from a failure in knowledge, but not a failure in scientific knowledge. Such a person will not understand why the death of a human is a tragedy but the explosion of a lifeless galaxy is merely a firework.

There is one final question. If the irreducible worth of persons is a fact, must the universe be substantially different from the picture that science alone gives of it? What must the universe contain, at a minimum, if its contents include not only stones, galaxies, atoms, and brains but also beings of moral worth?

That is an extremely difficult question. Science has taken us to the brink of another kind of knowledge, by opening up to us a vast range of natural facts but also exposing us to the limits of its own ways of knowing. We cannot believe that what science knows is all there is.

ENDNOTES

Preface

1. M. Cox et al., "Eighteenth and Nineteenth Century Dental Restoration, Treatment and Consequences in a British Nobleman," *British Dental Journal* 9 (2000): 593–6; effects described in M. Addy, "Dentine Hypersensitivity: New Perspectives on an Old Problem," *International Dental Journal* 52 (2002): 367–75.

Chapter 1—Evidence

1. Classic account in J. M. Keynes, *A Treatise on Probability* (London, 1921); recent full exposition in E. T. Jaynes, *Probability Theory: The Logic of Science* (Cambridge, 2003); introduction in D. C. Stove, *Probability and Hume's Inductive Scepticism* (Oxford, 1973), ch. 1.
2. P. Forrest, *The Dynamics of Belief: A Normative Logic* (Oxford, 1986), ch. 8; C. G. Hempel, *Aspects of Scientific Explanation* (New York, 1965), ch. 2.
3. G. Polya, *Patterns of Plausible Inference* (Princeton, 1968), 4.
4. D. C. Williams, *The Ground of Induction* (Cambridge, MA, 1947; New York, 1963); D. C. Stove, *The Rationality of Induction* (Oxford, 1986).

5. D. Stove, *Anything Goes: Origins of the Cult of Scientific Irrationalism* (Sydney, 1998), ch. 4.

6. J. B. Russell, *Inventing the Flat Earth: Columbus and Modern Historians* (New York, 1991).

7. J. J. Josephson and S. G. Josephson, eds., *Abductive Inference: Computation, Philosophy, Technology* (Cambridge, 1994); P. Lipton, *Inference to the Best Explanation*, 2nd ed. (London, 2004); L. Magnani, *Abduction, Reason and Science: Processes of Discovery and Explanation* (New York, 2001); Q. Nelson, *The Slightest Philosophy* (Indianapolis, 2007).

8. J. Franklin, *The Science of Conjecture: Evidence and Probability Before Pascal* (Baltimore, 2001), 348.

9. L. Rosa et al., "A Close Look at Therapeutic Touch," *Journal of the American Medical Association* 279 (1998): 1005–10; described in M. Gardner, *Are Universes Thicker than Blackberries?* (New York, 2003), ch. 26.

10. A. Franklin, *The Neglect of Experiment* (New York, 1986); further in A. Franklin, *Experiment, Right or Wrong* (New York, 1990).

Chapter 2—Enemies of Science: The Early Phase

1. D. C. Stove, *Popper and After: Four Modern Irrationalists* (Oxford, 1982); reprint, *Anything Goes: Origins of the Cult of Scientific Irrationalism* (Sydney, 1998).

2. Stove, *Anything Goes*, 103.

3. Stove, *Anything Goes,* 172–5.

4. K. Popper, *The Logic of Scientific Discovery* (London, 1959), 194–5, 197; Stove, *Anything Goes*, ch. 2.

5. J. C. Eccles, *Facing Reality* (New York, 1970), 108.

6. P. Medawar, BBC Radio 3 Broadcast, July 28, 1972, quoted in B. Magee, *Popper* (London, 1973), 9.

7. P. B. Medawar, *Induction and Intuition in Scientific Thought* (London, 1969), 51.

8. Galileo, *Dialogue Concerning the Two Chief World Systems*, 2nd ed., trans. S. Drake (Berkeley, 1967), 53–4.

9. I. Lakatos, "Falsification and the Methodology of Scientific Research Programmes," in *Criticism and the Growth of Knowledge*, eds. I. Lakatos and A. Musgrave (Cambridge, 1970), 164; Stove, *Anything Goes*, 32.

10. Critique in P. Slezak, "Sociology of Scientific Knowledge and Scientific Education: Part I," *Science and Education* 3 (1994): 265–94.

11. M. Charlesworth et al., *Life Among the Scientists* (Geelong, 1989).

Chapter 3—Enemies of Science: The Postmodernist Phase

1. N. Lucy, *Postmodern Literary Theory: An Introduction* (Oxford, 1997), 23, cf. 42; similar in D. Novitz, "The Rage for Deconstruction," *Monist* 69 (1986): 39–55.

2. R. Tallis, *Not Saussure: A Critique of Post-Saussurean Literary Theory* (Basingstoke, 1988); also R. Freadman and S. Miller, *Re-Thinking Theory* (Cambridge, 1992).

3. http://physics.nyu.edu/faculty/sokal/; further in A. Sokal, *Beyond the Hoax: Science, Philosophy and Culture* (Oxford, 2008).

4. D. C. Stove, "Judge's Report on the Competition to Find the Worst Argument in the World," in *Cricket Versus Republicanism*, ed. D. C. Stove (Sydney, 1995), 66–7; see J. Franklin, "Stove's Discovery of the Worst Argument in the World," *Philosophy* 77 (2002): 615–24.

5. D. C. Stove, *The Plato Cult and Other Philosophical Follies* (Oxford, 1991), 168.

6. Stove, *The Plato Cult*, 167.

7. A. Olding, "Religion as Smorgasbord," *Quadrant*, 42(5) (May, 1998): 73–5.

8. T. Kuhn, *The Structure of Scientific Revolutions*, 2nd ed. (Chicago, 1970), 206–7.

9. D. Bloor, *Knowledge and Social Imagery*, 2nd ed. (London, 1976), 171–2.

10. B. Barnes, D. Bloor, and J. Henry, *Scientific Knowledge: A Sociological Analysis* (Chicago, 1996), 48.

11. B. Barnes and D. Bloor, "Relativism, Rationalism and the Sociology of Knowledge," in *Rationality and Relativism*, ed. M. Hollis and S. Lukes (Oxford, 1981), 23.

12. Bloor, *Knowledge and Social Imagery*, 12.

13. Editor's introduction to A. Ross, *Science Wars* (Durham, NC, 1996), 14; replies from the other side, including Sokal's comments, in N. Koertge, ed., *A House Built on Sand: Exposing Post-Modern Myths About Science* (New York, 1998).

14. A. Sokal and J. Bricmont, *Intellectual Impostures: Postmodern Philosophers' Abuse of Science* (London, 1998); more analysis of the causes in P. R. Gross and N. Levitt, *The Higher Superstition: The Academic Left and Its Quarrels with Science* (Baltimore, 1994).

15. W. Bogard, "Sense and Segmentarity: Some Markers of Deleuzian-Guattarian Sociology," *Sociological Theory* 16 (1988): 53.

16. C. Colebrook, *Gilles Deleuze* (London, 2002), 3.

17. B. Massumi, *Parables for the Virtual: Movement, Affect, Sensation* (Durham, NC, 2002), 37; critique in P. Bell, "Neo-psychology or neo-humans? A critique of Massumi's 'Parables for the virtual,'" *Continuum: Journal of Media and Cultural Studies*, 17 (2003): 445–62.

Chapter 4—The Furniture

1. B. C. Smith, *On the Origin of Objects* (Cambridge, MA, 1996).

2. D. M. Armstrong, *A World of States of Affairs* (Cambridge, 1997).

3. T. Nagel, *The View from Nowhere* (New York, 1986).

4. R. Grossmann, *The Categorial Structure of the World* (Bloomington, IN, 1983).

5. A philosophical view in H. Chang, *Inventing Temperature: Measurement and Scientific Progress* (New York, 2004), ch. 4.

6. L. Corbesier and G. Coupland, "The Search for Florigen: A Review of Recent Progress," *Journal of Experimental Botany* 57 (2006): 3395–3403.

7. B. Chandrasekaran, J. J. Josephson, and V. R. Benjamins, "What Are Ontologies, and Why Do We Need Them?" *IEEE Intelligent*

Systems 14, no. 1 (Jan/Feb 1999): 20–26; B. Smith, "Ontology," in *Blackwell Guide to the Philosophy of Computing and Information*, ed. L. Floridi (Oxford, 2003), 155–66, preprint at http://ontology. Buffalo.edu/smith/articles/ontology_pic.pdf.

8.   J. L. Borges, "The Analytical Language of John Wilkins," in *Other Inquisitions 1937–1952*; one translation at http://www.crockford.com/wrrrld/wilkins.html.

9.   M. Foucault, *The Order of Things: An Archaeology of the Human Sciences* (London, 1970), preface; K. Windschuttle, "Absolutely Relative," *National Review* (Sept. 15, 1997), http://www.nationalreview.com/15sept97/windschuttle091597.html.

10.   C. B. Mervis and E. Rosch, "Categorization of Natural Objects," *Annual Reviews in Psychology* 32 (1981): 89–115.

11.   J. Franklin, "Mental Furniture from the Philosophers," *Et Cetera* 40 (1983): 177–91.

12.   C. Dilworth, *The Metaphysics of Science,* 2nd ed. (Dordrecht, 2006), ch. 7; D. S. Oderberg, *Real Essentialism* (New York, 2007), chs. 5, 9; history in A. C. Crombie, *Styles of Scientific Thinking in the European Tradition,* vol. 2 (London, 1994), chs. 15–16.

13.   A. H. Fielding, *Cluster and Classification Techniques for the Biosciences* (Cambridge, 2007).

14.   D. M. Armstrong, *What Is a Law of Nature?* (Cambridge, 1983).

15.   S. Mumford, *Dispositions* (Oxford, 1998); G. Molnar, *Powers: A Study in Metaphysics* (Oxford, 2003).

16.   P. E. Gibbs, "The Small Scale Structure of Space-Time: A Bibliographical Review (1995)," http://arxiv.org/abs/hep-th/9506171.

17.   G. Nerlich, *The Shape of Space,* 2nd ed. (Cambridge, 1994).

18.   I. Newton, "Scholium to the Definitions," in *Philosophiae Naturalis Principia Mathematica,* bk. 1 (1689).

19.   H. Price, *Time's Arrow and Archimedes' Point* (New York, 1996).

20.   W. Sellars, "Philosophy and the Scientific Image of Man," in *Science, Perception and Reality,* ed. W. Sellars (London, 1963), http://www.ditext.com/sellars/psim.html.

21.   J. Franklin, *The Science of Conjecture: Evidence and Probability before Pascal* (Baltimore, 2001), 179.

22. C. L. Hardin, *Color for Philosophers: Unweaving the Rainbow* (Indianapolis, 1988); A. Byrne and D. R. Hilbert, eds., *Readings on Color* (Cambridge, MA, 1997).

23. E. Grant, *A Source Book in Medieval Science* (Cambridge, MA, 1974), sections 42–3.

Chapter 5—The Physical Sciences

1. G. N. Derry, *What Science Is: And How It Works* (Princeton, 1999), 42–3.

2. Galileo, *Dialogue Concerning the Two Chief World Systems*, trans. Stillman Drake (Berkeley, 1953), 186–7.

3. Galileo, *Dialogue*, Second Day (Drake, 190–1); see W. R. Shea, *Galileo's Intellectual Revolution* (London, 1972), 154–5; R. Sorenson, *Thought Experiments* (New York, 1992), 88–92.

4. M. McCloskey, "Intuitive Physics," *Scientific American* 248, no. 4 (Apr. 1983): 114–22; M. K. Kaiser, M. McCloskey, and D. R. Proffitt, "Development of Intuitive Theories of Motion: Curvilinear Motion in the Absence of External Forces," *Developmental Psychology* 22 (1986): 67–71.

5. J. R. Brown, *The Laboratory of the Mind: Thought Experiments in the Natural Sciences* (London, 1991), 31; J. Lennox, "Darwinian Thought Experiments: A Function for Just-So Stories," in *Thought Experiments in Science and Philosophy*, eds. T. Horowitz and G. Massey (Savage, MD, 1991), 223–46.

6. J. Honner, "The Transcendental Philosophy of Niels Bohr," *Studies in History and Philosophy of Science A* 13 (1982): 1–29; W. Heisenberg, *Physics and Beyond* (New York, 1971), 13, 122–3.

7. D. J. Griffiths, *Introduction to Quantum Mechanics*, 2nd ed. (Upper Saddle River, NJ, 2005), ch. 1.

8. Griffiths, *Introduction to Quantum Mechanics*, 431.

9. M. S. Turner, "A Sober Assessment of Cosmology at the New Millennium," *Publications of the Astronomical Society of the Pacific* 113 (2001): 653–7.

10. I. Nicolson, *Dark Side of the Universe: Dark Matter, Dark Energy and the Fate of the Cosmos* (Baltimore, 2007).

Chapter 6—Biology and Cognition

1. M. G. Frank and J. H. Benington, "The Role of Sleep in Memory Consolidation and Brain Plasticity: Dream or Reality? *The Neuroscientist* 12 (2006): 477–88.

2. J. M. Siegel, "Do all Animals Sleep?" *Trends in Neurosciences* 31 (2008): 208–13.

3. A. M. Turing, "Computing Machinery and Intelligence," *Mind* 59 (1950): 433–60, available at http://loebner.net/Prizef/TuringArticle.html.

4. H. A. Simon and A. Newell, "Heuristic Problem Solving: The Next Advance in Operations Research," *Operations Research* 6 (1958): 1–10.

5. S. Derbyshire, "We're No Slaves to Our Senses," *Spiked Review of Books* 4 (Aug 2007).

6. R. Shepard and J. Metzler, "Mental Rotation of Three-Dimensional Objects," *Science* 171, no. 972 (1971), 701–3; R. Shepard and L. Cooper, *Mental Images and Their Transformations* (Cambridge, MA, 1982).

7. H. Gardner, *The Mind's New Science: A History of the Cognitive Revolution* (New York, 1985).

8. S. Z. Li and A. K. Jain, eds., *Handbook of Face Recognition* (New York, 2005).

9. D. Marr, *Vision: A Computational Investigation into the Human Representation and Processing of Visual Information* (San Francisco, 1982); E. R. Davies, *Machine Vision: Theory, Algorithms, Practicalities*, 3rd ed. (Amsterdam, 2005).

10. A. Gopnik, *The Philosophical Baby: What Children's Minds Tell Us About Truth, Love, and the Meaning of Life* (New York, 2009).

11. A. Gopnik, A. Meltzoff, and P. Kuhl, *How Babies Think: The Science of Childhood* (London, 1999).

12. L. E. Bahrick, R. Lickliter, and R. Flom, "Intersensory Redundancy Guides the Development of Selective Attention, Perception, and Cognition in Infancy," *Current Directions in Psychological Science* 13 (2004): 99–102; M. A. Schmuckler, "Visual-Proprioceptive Intermodal Perception in Infancy," *Infant Behavior and Development* 19 (1996): 221–32.

13. Gopnik et al., ch. 5.

Chapter 7—Mathematics

1. B. Russell, "Recent Work on the Principles of Mathematics, 1901" in *Collected Papers of Bertrand Russell*, vol. 3, ed. G. H. Moore (London and New York, 1993), 366.
2. A. Einstein, *Ideas and Opinions* (New York, 1954), 233.
3. H. Weyl, *Symmetry* (Princeton, 1952); G. Hon and B. R. Goldstein, *From Summetria to Symmetry: The Making of a Revolutionary Scientific Concept* (Dordrecht, 2008).
4. I. Newton, *Arithmetica Universalis* (1728), 2; discussion in J. Bigelow and R. Pargetter, *Science and Necessity* (Cambridge, 1990), ch. 2.
5. B. Ellis, *Basic Concepts of Measurement* (Cambridge, 1968), ch. 4; J. Michell, "Numbers as Quantitative Relations and the Traditional Theory of Measurement," *British Journal for the Philosophy of Science* 45 (1994): 389–406.
6. J. Franklin and A. Daoud, *Proof in Mathematics: An Introduction* (Sydney, 1996), chs. 1–2.
7. Many examples in J. R. Brown, *Philosophy of Mathematics: An Introduction to the World of Proofs and Pictures* (London, 1999).
8. J. L. Casti, *Mathematical Mountaintops: The Five Most Famous Problems of all Time* (New York, 2001).
9. E. W. Weisstein, "Kepler Conjecture, from MathWorld—A Wolfram Web Resource," http://mathworld.wolfram.com/Kepler-Conjecture.html.
10. D. H. Fowler, *The Mathematics of Plato's Academy: A New Reconstruction* (Oxford, 1987).
11. D. J. Fairbanks and B. Rytting, "Mendelian Controversies: A Botanical and Historical Review," *American Journal of Botany* 88 (2001): 737–52; E. Novitski, "On Fisher's Criticism of Mendel's Results with the Garden Pea," *Genetics* 166 (2004): 1133–6.
12. R. B. Banks, *Towing Icebergs, Falling Dominoes, and Other Adventures in Applied Mathematics* (Princeton, 1998); F. Giordano, M. D. Weir, and W. P. Fox, *A First Course in Mathematical*

Modeling, 3rd ed. (Pacific Grove, CA, 2003); many examples in the "COMAP Mathematical Contest in Modeling," http://www.comap.com/undergraduate/contests/.

13. P. M. Allen et al., *Models of Urban Settlement and Structure as Dynamic Self-Organizing Systems* (Washington, DC, 1981); I. Benenson and P. N. Torrens, *Geosimulation: Automata-based Modelling of Urban Phenomena* (Hoboken, NJ, 2004).

14. http://esa.un.org/unpp/.

15. N. Biggs, E. Lloyd, and R. Wilson, *Graph Theory: 1736–1936* (Oxford, 1976), 3–8.

16. A 2D animation of heat diffusion at http://arantxa.ii.uam.es/~jlara/investigacion/ecomm/pdes/CA.html.

17. L. F. Richardson, *Weather Prediction by Numerical Process*, 2nd ed. (London, 1922; Cambridge, 2007); P. Lynch, *The Emergence of Numerical Weather Prediction: Richardson's Dream* (Cambridge, 2006).

Chapter 8—Enemies of Mathematics

1. I. Lakatos, *Proofs and Refutations: The Logic of Mathematical Discovery* (Cambridge, 1976); discussion in B. Larvor, *Lakatos: An Introduction* (London, 1998), ch. 2.

2. http://mathforum.org/kb/plaintext.jspa?messageID=1375599; discussion in G. Hanna, "The Ongoing Value of Proof," in *Theorems in School: From History, Epistemology and Cognition to Classroom Practice*, ed. P. Boero (Rotterdam, 2007), 3–16.

3. D. C. Stove, *Anything Goes: Origins of the Cult of Scientific Irrationalism* (Sydney, 1998), 32.

4. R. Debray, *Le Scribe* (Paris, 1980), 70; quoted in A. Sokal and J. Bricmont, *Intellectual Impostures* (London, 1998), 167.

5. G. Deleuze and F. Guattari, *What Is Philosophy?* (New York, 1994), 122, in *Intellectual Impostures* 150–1; a Deleuzian book on "mathematics" S. Duffy, ed., *Virtual Mathematics: The Logic of Difference* (Manchester, 2006); Australian imitations in J. Franklin, *Corrupting the Youth: A History of Philosophy in Australia* (Sydney, 2003), 368.

6. P. Ernest, *Social Constructivism as a Philosophy of Mathematics* (Albany NY, 1998), 26.
7. M. Kline, *Mathematics: The Loss of Certainty* (New York, 1980).
8. Gross and Levitt, *Higher Superstition*, 113–5.
9. E. de Freitas, "Plotting Intersections Along the Political Axis: The Interior Voice of Dissenting Mathematics Teachers," *Educational Studies in Mathematics* 55 (2004): 259–74.

Chapter 9—The Formal Sciences

1. M. M. Waldrop, *Complexity: The Emerging Science at the Edge of Order and Chaos* (New York, 1992); P. Bak, *How Nature Works: The Science of Self-Organized Criticality* (Oxford, 1997); S. Wolfram, *A New Kind of Science* (Champaign, IL, 2002); H. A. Simon, *The Sciences of the Artificial*, 2nd ed. (Cambridge, MA, 1969; 1981).
2. F. N. Trefethen, "History of Operations Research," in *Operations Research for Management*, vol. 1, eds. J. F. McCloskey and F. N. Trefethen (Baltimore, 1954), 3–35; S. I. Gass and A. A. Assad, eds., *An Annotated Timeline of Operations Research: An Informal History* (New York, 2004).
3. C. H. Waddington, *O. R. in World War 2: Operational Research Against the U-boat* (London, 1973), ch. 7.
4. C. D. Woolsey and H. S. Swanson, *Operations Research for Immediate Application: A Quick and Dirty Manual* (New York, 1975); H. A. Taha, *Operations Research: An Introduction*, 7th ed. (Upper Saddle River, NJ, 2003).
5. R. C. Dorf and R. H. Bishop, *Modern Control Systems*, 10th ed. (Upper Saddle River, NJ, 2004).
6. Waldrop, *Complexity*, 63.
7. http://web.onetel.net.uk/~hibou/Beer%20and%20Nappies.html.
8. I. Ayres, *Super Crunchers: How Anything Can Be Predicted* (London, 2007).
9. D. Jungnickel, *Graphs, Networks and Algorithms* (Berlin, 2005); M. G. H. Bell and Y. Iida, *Transportation Network Analysis*

(Chichester, 1997); J. F. Hayes and T. V. J. G. Babu, *Modeling and Analysis of Telecommunications Networks* (Hoboken, NJ, 2004).

10. B. D. Bunday, *Basic Queueing Theory* (London, 1986).

11. C. E. Shannon, *The Mathematical Theory of Communication* (Urbana, IL, 1949); D. R. Hankerson, G. A. Harris, and P. D. Johnson, *Introduction to Information Theory and Data Compression*, 2nd ed. (Boca Raton, FL, 2003).

12. M. S. Waterman, ed., *Mathematical Models for DNA Sequences* (Boca Raton, FL, 1989); J. Augen, *Bioinformatics in the Post-Genomic Era* (Boston, 2005); A. Datta and E. R. Dougherty, *Introduction to Genomic Signal Processing with Control* (Boca Raton, FL, 2007).

13. J. von Neumann and O. Morgenstern, *Theory of Games and Economic Behavior* (Princeton, 1944); L. G. Telser, *Competition, Collusion and Game Theory* (London, 1972); M. Shubik, *Game Theory in the Social Sciences* (Cambridge, MA, 1982); A. K. Dixit and S. Skeath, *Games of Strategy*, 2nd ed. (New York, 2004).

14. M. Enquist, *Game Theory Studies on Aggressive Behaviour* (Stockholm, 1984).

15. M. Gardner, "Mathematical Games: The Fantastic Combinations of John Conway's New Solitaire Game 'Life,'" *Scientific American* 223, no. 4 (1970): 120–3; W. Poundstone, *The Recursive Universe* (Oxford, 1987), 24–47 and chs. 4, 6, 8, 10; C. Adami, *Introduction to Artificial Life* (New York, 1998).

16. Animations in Wikipedia article "Conway's Game of Life," http://en.wikipedia.org/wiki/Conway's_Game_of_Life.

17. R. May, *Theoretical Ecology: Principles and Applications*, 2nd ed. (Oxford, 1981); M. Kot, *Elements of Mathematical Ecology* (Cambridge, 2001).

18. P. C. W. Davies, *The Cosmic Blueprint* (New York, 1988); E. Jantsch, *Self-Organizing Universe* (Oxford, 1980); S. A. Kauffman, *At Home in the Universe: The Search for Laws of Self-Organization and Complexity* (New York, 1995); Wolfram, *A New Kind of Science*, chs. 7–9.

19. H. S. Wilf, *Algorithms and Complexity* (Englewood Cliffs, NJ, 1986); L. Fortnow and S. Homer, "A Short History of Complexity

Theory," in *Handbook of the History of Mathematical Logic,* eds. D. van Dalen, J. Dawson, and A. Kanamori; available at http://people.cs.uchicago.edu/~fortnow/papers/history.pdf.

20. P. Frey, ed., *Chess Skill in Man and Machine,* 2nd ed. (New York, 1983); F.-H. Hsu, *Behind Deep Blue* (Princeton, NJ, 2002).

21. B. Boehm, "A View of 20th and 21st Century Software Engineering," *Proceedings of the 28th International Conference on Software Engineering* (May 2006): 12–29.

22. C. A. R. Hoare, "An Axiomatic Basis for Computer Programming," *Communications of the Association for Computing Machinery* 12 (1969): 576–80.

23. R. Stallman and S. Garfinkle, "Against Software Patents," *Communications of the Association for Computing Machinery* 35 (1992), 17–22, 121.

24. J. H. Fetzer, "Program Verification: The Very Idea," *Communications of the Association for Computing Machinery* 31 (1988), 1048–63.

25. J. H. Fetzer, "Program Verification Reprise: The Author's Response," *Communications of the Association for Computing Machinery* 32 (1989): 377–81; further in J. H. Fetzer, "Philosophy and Computer Science: Reflections on the Program Verification Debate," in *The Digital Phoenix,* ed. T. W. Bynum (Cambridge, 1998), 253–73.

26. J. Barwise, "Mathematical Proofs of Computer System Correctness," *Notices of the American Mathematical Society* 36 (1989): 844–51.

27. M. M. Lehman, "Uncertainty in Computer Application," *Communications of the Association for Computing Machinery* 33 (1990): 584–6.

28. D. Gries, *The Science of Programming* (New York, 1981), 5.

29. C. A. R. Hoare, "Programs are Predicates," in *Mathematical Logic and Programming Languages,* eds. C. A. R. Hoare and J. C. Stephenson (Englewood Cliffs, NJ, 1985), 141–55.

30. With the exception of David Hume: *Treatise of Human Nature,* I. IV.i.

31. H. M. Müller, letter in *Communications of the Association for Computing Machinery* 32 (1989): 506–8.

32. J. Dobson and B. Randell, "Program Verification: Public Image and Private Reality," *Communications of the Association for Computing Machinery* 32 (1989): 420–2; D. Mackenzie, "The Fangs of the VIPER," *Nature* 352 (1991): 467–8; later developments in C. Kern and M. R. Greenstreet, "Formal Verification in Hardware Design: A Survey," *ACM Transactions on Design Automation of Electronic Systems* 4 (1999): 123–93.

33. D. Mackenzie, "Computers, Formal Proofs and the Law Courts," *Notices of the American Mathematical Society* 39 (1992): 1066–9.

Chapter 10—Probabilities and Risks

1. Introductions in I. Hacking, *The Emergence of Probability*, 2nd ed. (Cambridge, 2006), ch. 2; D. C. Stove, *Probability and Hume's Inductive Scepticism* (Oxford, 1973), ch. 1; D. H. Mellor, *Probability: A Philosophical Introduction* (London, 2005), ch. 1.

2. D. Kahneman, P. Slovic, and A. Tversky, *Judgment Under Uncertainty: Heuristics and Biases* (Cambridge, 1982).

3. J. Franklin, "Case comment: *United States* v. *Copeland*, 369 F. Supp. 2nd 275 (E. D. N. Y. 2005): quantification of the 'proof beyond reasonable doubt' standard," *Law, Probability and Risk* 5 (2006), 159–65.

4. Reviews in M. Strevens, "A Closer Look at the New Principle," *British Journal for the Philosophy of Science* 46 (1995): 545–61; B. Loewer, "David Lewis's Humean Theory of Objective Chance," *Philosophy of Science* 71 (2004), 1115–25.

5. J. Franklin, "Non-Deductive Logic in Mathematics," *British Journal for the Philosophy of Science* 38 (1987): 1–18; many examples in G. Polya, *Mathematics and Plausible Reasoning*, 2 vols. (London, 1954).

6. D. Knuth, *The Art of Computer Programming*, vol. 2, 3rd ed. (Reading, MA, 1997), ch. 3.

7. R. A. Fisher, "The Expansion of Statistics," *J. of the Royal Statistical Society* A 116 (1953): 1–6.

8. R. A. Fisher, *The Design of Experiments* 7th ed., (Edinburgh, 1935, 1960), ch. 2.

9. D. Salsburg, *The Lady Tasting Tea: How Statistics Revolutionized Science in the Twentieth Century* (New York, 2001), 8; but according to J. F. Box, *R. A. Fisher: The Life of a Scientist* (New York, 1978), the venue was Rothamsted and the lady, Muriel Bristol.

10. *Talmud*, Yebamot 64b, quoted and discussed in J. Franklin, *The Science of Conjecture: Evidence and Probability Before Pascal* (Baltimore, 2001), 174.

11. G. Gigerenzer, *Gut Feelings: The Intelligence of the Unconscious* (New York, 2007).

12. G. Gigerenzer and D. J. Murray, *Cognition as Intuitive Statistics* (Hillsdale, NJ, 1987).

13. Franklin, *The Science of Conjecture*, 324–5.

14. C. Yu and L. Smith, "Rapid Word Learning Under Uncertainty Via Cross-Situational Statistics," *Psychological Science* 18 (2007): 414–20; E. Téglás, V. Girotto, M. Gonzalez, and L. L. Bonatti, "Intuitions of Probabilities Shape Expectations About the Future at 12 Months and Beyond," *Proceedings of the National Academy of Sciences* 104 (2007): 19156–9.

15. J. Franklin, "Risk-Driven Global Compliance Regimes in Banking and Accounting: The New Law Merchant," *Law, Probability and Risk* 4 (2005): 237–50.

16. J. Franklin, S. A. Sisson, M. A. Burgman, and J. K. Martin, "Evaluating Extreme Risks in Invasion Ecology: Learning from Banking Compliance," *Diversity and Distributions* 14 (2008): 581–91.

17. Salsburg, *The Lady Tasting Tea*, ch. 6; S. Coles, *An Introduction to Statistical Modeling of Extreme Values* (London, 2001).

18. Franklin, Sisson et al., appendix.

19. E. F. Fama, "The Behavior of Stock-Market Prices," *Journal of Business* 38 (1965): 34–105.

20. "Transcript: Defense Department Briefing," February 12, 2002, http://www.globalsecurity.org/military/library/news/2002/02/mil-020212-usia01.htm.

21. United States Nuclear Regulatory Commission, "Backgrounder on Fire Protection for Nuclear Power Plants," (2006), http://www.nrc.gov/reading-rm/doc-collections/fact-sheets/fire-pro-

tection-bg.html; other examples in N. Taleb, *The Black Swan: The Impact of the Highly Improbable* (New York, 2007).

Chapter 11—Are the Social Sciences *Sciences*?

1. J. Buchan, review of P. Ball, "Critical Mass," *The Guardian*, March 27 2004, http://books.guardian.co.uk/reviews/politics philosophyandsociety/0,6121,1178982,00.html.

2. A. Quetelet, *A Treatise on Man, and the Development of His Faculties,* trans. S. Diamond (Edinburgh, 1842), 80.

3. M. Schreckenberg and R. Selten, eds., *Human Behaviour and Traffic Networks* (Berlin, 2004); summary in P. Ball, *Critical Mass: How One Thing Leads to Another* (London, 2004), ch. 7.

4. E.g. C. Marrison, *The Fundamentals of Risk Measurement* (New York, 2002).

5. R. Lowenstein, *When Genius Failed: The Rise and Fall of Long-Term Capital Management* (New York 2000).

6. F. A. Hayek, *Law, Legislation and Liberty: The Mirage of Social Justice,* vol. 2 (London, 1976).

7. D. A. Lawlor, G. D. Smith, and S. Ebrahim, "The Hormone Replacement-Coronary Heart Disease Conundrum: Is this the Death of Observational Epidemiology?" *International Journal of Epidemiology* 33 (2004): 464–7.

8. R. Brown, *Rules and Laws in Sociology* (London, 1973).

9. A. Gopnik and L. Schulz, *Causal Learning: Psychology, Philosophy, and Computation* (Oxford, 2007), editors' introduction, 3–4.

10. R. E. Neapolitan, *Learning Bayesian Networks* (Harlow, 2004).

11. A scientific perspective in M. Buchanan, *Ubiquity: The Science of History* (London, 2000).

12. K. Windschuttle, *The Killing of History* (New York, 1997), ch. 7; C. B. McCullagh, *The Truth of History* (London, 1998).

13. A. Gopnik, A. Meltzoff, and P. Kuhl, *How Babies Think: The Science of Childhood* (London, 1999), 32–3.

14. S. E. Gleason and W. L. Langer *The Undeclared War, 1940–1941* (1953), 533.

15. J. Lukacs, *June 1941: Hitler and Stalin* (New Haven, CT, 2006); D. E. Murphy, *What Stalin Knew: The Enigma of Barbarossa* (New Haven, CT, 2005); C. Pleshakov, *Stalin's Folly: The Tragic First Ten Days of World War Two on the Eastern Front* (Boston, 2005).

Chapter 12—Actually Existing Science: Institutions for Knowing

1. G. J. Feist, *The Psychology of Science and the Origins of the Scientific Mind* (New Haven, 2006), 23, 72–3, 116, 122; portrait of a stereotype in J. Gleick, *Genius: The Life and Science of Richard Feynman* (New York, 1992).

2. Feist, 118–9; S. Baron-Cohen et al., "The Autism-Spectrum Quotient (AQ): Evidence from Asperger Syndrome/High-Functioning Autism, Males and Females, Scientists and Mathematicians," *Journal of Autism and Developmental Disorders* 31 (2001): 5–17.

3. D. Kuhn and S. Pearsall, "Developmental Origins of Scientific Thinking," *Journal of Cognition and Development* 1 (2000), 113–29; Feist, 64–7.

4. http://www.gatesfoundation.org/GlobalHealth/Pri_Diseases/Malaria/.

5. J. Gillies and R. Cailliau, *How the Web Was Born: The Story of the World Wide Web* (Oxford, 2000).

6. D. Sherry, "On Mathematical Error," *Studies in History and Philosophy of Science A* 28 (1997): 393–416; 476 errors up to 1916, many trivial, listed in M. Lecat, *Erreurs de mathematicians des origines à nos jours* (Brussels, 1935).

7. A. Yankauer, "Who Are the Peer Reviewers and How Much Do They Review?" *Journal of the American Medical Association* 263 (1990): 1338–40.

8. R. Horton, "Genetically Modified Food: Consternation, Confusion, and Crack-up," *Medical Journal of Australia* 177 (2000): 148–9.

9. G. Edmond, "Judging the Scientific and Medical Literature: Some Legal Implications of Changes to Biomedical Research

and Publication," *Oxford Journal of Legal Studies* 28 (2008): 523–61, with reference to W. Broad and N. Wade, *Betrayers of the Truth: Fraud and Deceit in the Halls of Science* (New York, 1982), 79–82, 203–11.

10. A. T. Winfree, "The Prehistory of the Belousov-Zhabotinsky Oscillator," *Journal of Chemical Education* 61 (1984): 661–3.

11. D. Witztum, E. Rips, and Y. Rosenberg, "Equidistant Letter Sequences in the Book of Genesis," *Statistical Science* 9 (1994), 429–38.

12. M. Drosnin, *The Bible Code* (New York, 1997).

13. B. McKay, D. Bar-Natan, M. Bar-Hillel, and G. Kalai, "Solving the Bible Code Puzzle," *Statistical Science* 14 (1999):150–73; A. M. Hasofer, "A Statistical Critique of the Witztum *et al* Paper" (February 18, 1998), Expert Opinions on the Bible Codes. http://cs.anu.edu.au/~bdm/dilugim/opinions/hasofer_s.pdf.

14. K. R. Foster and P. W. Huber, *Judging Science: Scientific Knowledge and the Federal Courts* (Cambridge, MA, 1997).

15. P. J. Weindling, *Nazi Medicine and the Nuremberg Trials: From Medical War Crimes to Informed Consent* (Basingstoke, 2004); N. Baumslag, *Murderous Medicine: Nazi Doctors, Human Experimentation, and Typhus* (Westport, 2005); D. Barenblatt, *A Plague upon Humanity: The Secret Genocide of Axis Japan's Germ Warfare Operation* (New York, 2004).

16. A. Collins, *In the Sleep Room: The Story of CIA Brainwashing Experiments in Canada* (Toronto, 1997); G. Thomas, *Journey into Madness: The True Story of Secret CIA Mind Control and Medical Abuse* (New York, 1989).

17. C. Paul, "Internal and External Morality of Medicine: Lessons from New Zealand," *British Medical Journal* 320 (2000): 499–503.

Chapter 13—The Complexity Obstacle to Knowledge

1. E.g. D. Penny, "An Interpretive Review of the Origin of Life Research," *Biology and Philosophy* 20 (2005): 633–671.

2. National Academy of Sciences and The Institute of Medicine, *Science, Evolution and Creationism* summary brochure (Wash-

ington DC, 2008), 2. http://www.nap.edu/catalog.php?record_id
=11876.

3. D. C. Stove, *Darwinian Fairytales* (New York, 2006).

4. G. de Beer, *Homology: An Unsolved Problem* (Oxford, 1973); J. R. Hinchliffe, "The Developmental Basis of Limb Evolution," *International Journal of Developmental Biology* 46 (2002), 835–45.

5. L. A. Pray, "Epigenetics: Genome, Meet Your Environment," *The Scientist* 18, no. 13 (July 5, 2004): 14–20.

6. L. Huxley, ed., *The Life and Letters of Thomas Henry Huxley* (London, 1900), vol. 1, ch. 1.13.

7. R. Dawkins, "Why Darwin Matters," *Guardian* Feb 9, 2008, http://www.guardian.co.uk/science/2008/feb/09/darwin.dawkins1.

8. M. Kimura, "Natural Selection as the Process of Accumulating Genetic Information in Adaptive Evolution," *Genetical Research* 2 (1961): 127–40.

9. C. Darwin, *On the Origin of Species,* 1st ed. (London, 1959), 189.

10. M. J. Behe, *Darwin's Black Box: The Biochemical Challenge to Evolution* (New York, 1996, 2005); M. J. Behe, *The Edge of Evolution: The Search for the Limits of Darwinism* (New York, 2007).

11. M. Behe, "Evidence for Intelligent Design from Biochemistry," (a speech given at the Discovery Institute's God & Culture Conference, Seattle, 1996), http://www.arn.org/docs/behe/mb_idfrombiochemistry.htm.

12. K. R. Miller, "The Flagellum Unspun: The Collapse of Irreducible Complexity," in *Debating Design: from Darwin to DNA,* eds. M. Ruse and W. Dembski (New York, 2004).

13. M. J. Pallen and N. J. Matzke, "From the *Origin of Species* to the Origin of Bacterial Flagella," *Nature Reviews Microbiology* (Sept. 5, 2006).

14. E.g. W. R. Kininmonth, *Climate Change: A Natural Hazard* (Brentwood, 2004).

15. M. Wild et al., "From Dimming to Brightening: Decadal Changes in Solar Radiation at Earth's Surface," *Science* 308 (2005): 847–850; C. Ruckstuhl et al., "Aerosol and Cloud Effects

on Solar Brightening and the Recent Rapid Warming," *Geophysical Research Letters* 35 (2008): L12708.

16. D. W. J. Thompson et al., "A Large Discontinuity in the Mid-Twentieth Century in Observed Global-Mean Surface Temperature," *Nature* 453 (May 29, 2008): 646–9.

17. W. S. Broecker, "Paleoclimate: Was the Medieval Warm Period Global?" *Science* 291 (2001): 1497–9; E. Jansen et al. "Palaeoclimate" in *Climate Change 2007: The Physical Science Basis. Contribution of Working Group I to the Fourth Assessment Report of the IPCC*, ed. S. Solomon et al. (2007): 468–9, http://www.ipcc.ch/pdf/assessment-report/ar4/wg1/ar4-wg1-chapter6.pdf.

18. I. Plimer, *Heaven and Earth: Global Warming, The Missing Science* (Ballan, 2009).

19. K. L. Denman et al. "Couplings Between Changes in the Climate System and Biogeochemistry," in *Climate Change 2007*, section 7.5, http://www.ipcc.ch/pdf/assessment-report/ar4/wg1/ar4-wg1-chapter7.pdf; comments on the performance of models in general in D. Koutsoyiannis, "On the Credibility of Climate Predictions," *Hydrological Sciences Journal* 53 (2008): 671–84.

20. E.g. C. Jones et al., "Global Climate Change and Soil Carbon Stocks: Predictions from Two Contrasting Models for the Turnover of Organic Carbon in Soil," *Global Change Biology* 11 (2004): 154–66.

21. S. Sitch et al., "Assessing the Carbon Balance of Circumpolar Arctic Tundra Using Remote Sensing and Process Modeling," *Ecological Applications* 17 (2007): 213–34.

22. J. R. Petit et al., "Climate and Atmospheric History of the Past 420,000 Years from the Vostok Ice Core, Antarctica," *Nature* 399 (1999): 429–36.

23. E. Jansen et al. "Palaeoclimate," in *Climate Change 2007*, section 6.4, 446. http://www.ipcc.ch/pdf/assessment-report/ar4/wg1/ar4-wg1-chapter6.pdf.

24. K. Ravilious, "Mars Melt Hints at Solar, Not Human, Cause for Warming, Scientist Says," *National Geographic News*, Feb. 28, 2007, http://news.nationalgeographic.com/news/2007/02/070228-mars-warming.html.

25. K. Than, "Dust Storms Fuel Global Warming on Mars," Space.com, Apr. 4, 2007, http://www.space.com/scienceastronomy/070404_gw_mars.html.
26. http://climatedebatedaily.com/.

Chapter 14—Is That All There Is?

1. R. C. Stalnaker, "What Is It like to Be a Zombie?" in *Ways a World Might Be: Metaphysical and Anti-Metaphysical Essays* (Oxford, 2003), 239–53; D. Chalmers, "Consciousness and Its Place in Nature," in *Blackwell Guide to Philosophy of Mind*, ed. S. P. Stich and T. A. Warfield (Malden, MA, 2003), ch. 5; J. Shear, ed. *Explaining Consciousness—The "Hard Problem"* (Cambridge, MA, 1997); objections in D. Dennett, *Sweet Dreams: Philosophical Obstacles to a Science of Consciousness* (Cambridge, MA, 2005), ch. 1.
2. T. Nagel, "What Is It like to Be a Bat?" *Philosophical Review* (1974): 436.
3. C. McGinn, "Consciousness and Cosmology: Hyperdualism Ventilated," in *Consciousness*, ed. M. Davies and G. W. Humphreys (Oxford, 1993), 160.
4. D. M. Armstrong, "The Headless Woman Illusion and the Defence of Materialism," *Analysis* 29 (1968): 48–9.
5. "The Human Brain and Consciousness: In Conversation with Susan Greenfield," in *What Scientists Think*, ed. J. Stangroom (Abingdon, 2005), ch. 3.
6. S. Greenfield, *Journey to the Centers of the Mind: Toward a Science of Consciousness* (New York, 1995), ch. 4; S. Blackmore, *Consciousness: An Introduction* (Oxford, 2004), ch. 18.
7. B. Libet, "Unconscious Cerebral Initiative and the Role of Conscious Will in Voluntary Action," *Behavioral and Brain Sciences* 8 (1985): 529–66; B. Libet, "The Timing of Mental Events: Libet's Experimental Findings and Their Implications," *Consciousness and Cognition* 11 (2002): 291–99.
8. J. L. Mackie, *Ethics: Inventing Right and Wrong* (Harmondsworth, 1977).

9. S. Hawking, Master of the Universe, BBC TV, 1989, quoted in M. Shermer, *How We Believe: The Search for God in an Age of Science* (New York, 2000), 102; an older version in A. R. Wallace, "Man's Place in the Universe," *The Independent* 55 (1903): 473.

10. R. Dawkins, *The God Delusion* (London, 2006), 11–12.

11. T. Zerjal et al., "The Genetic Legacy of the Mongols," *American Journal of Human Genetics* 72 (2003): 717–21.

12. E. O. Wilson, *Sociobiology: The New Synthesis* (Cambridge, MA, 1975), 3.

13. J. Franklin, "Stove's Discovery of the Worst Argument in the World," *Philosophy* 77 (2002): 615–24.

14. R. Wright, *The Moral Animal: Evolutionary Psychology and Everyday Life* (New York, 1995); J. L. Mackie, "The Law of the Jungle," *Philosophy* 53 (1978): 455–64.

15. S. Milgram, *Obedience to Authority: An Experimental View* (London, 1974).

16. Stanford Prison Experiment site, http://www.prisonexp.org/.

17. M. D. Hauser, *Moral Minds: How Nature Designed Our Universal Sense of Right and Wrong* (New York, 2006).

18. R. Gaita, *Good and Evil: An Absolute Conception,* 2nd ed. (Abingdon, 2004), 315.

IMAGE CREDITS

pg. 12:

"Black Swan," watercolor by Richard Browne for Thomas Skottowe's manuscript *Select Specimens from Nature of the Birds, Animals, &c. &c. of New South Wales* (1813). Reprinted with permission from Mitchell Library, State Library of New South Wales.

pg. 19, FIGURE 1.1:

Reproduced from the Wikimedia Commons file "File: Aristarchus working.jpg." In the public domain per http://en.wikipedia.org/wiki/File:Aristarchus_working.jpg. Original Source: Library of Congress Vatican Exhibit, "Greek Mathematics and Its Modern Heirs." Vat. gr. 204 fol. 116 recto math06 NS.02. http://www.ibiblio.org/expo/vatican.exhibit/exhibit/d-mathematics/images/math06.jpg.

pg. 59, FIGURE 4.1:

Reprinted by permission from the publisher from "OpenCyc Selected Vocabulary and Upper Ontology." © 1996 by Cycorp, Inc. http://www.cyc.com/cycdoc/upperont-diagram.html.

pg. 77:
Sijp, Willem P. and Matthew H. England. "Effect of the Drake Passage Throughflow on Global Climate." *Journal of Physical Oceanography* 34 (2004): 1254–66. Courtesy of the authors.

pg. 83, FIGURE 5.1:
Reproduced from the Wikimedia Commons file "File: StevinEquilibrium.svg." In the public domain per http://en.wikipedia.org/wiki/File:StevinEquilibrium.svg. Original Source: http://www.dbnl.org/tekst/berk003voet01_01/berk003voet01_01_0002.htm.

pg. 101, FIGURE 6.1:
Shepard, Roger N. and Jacqueline Metzler. "Mental Rotation of Three-Dimensional Objects." *Science* 171 (1971): 701–3. Reprinted with permission from AAS.

pg. 185, FIGURE 11.1:
Adapted from the Wikimedia Commons file "Datei:Rl1000 cars300 p015 VDR.png." http://de.wikipedia.org/w/index.php?title=Datei:Rl1000_cars300_p015_VDR.png&filetimestamp=20050603224636 (May 25, 2009). Originally published by the GNU License for Free Documentation.

pg. 227, FIGURE 13.1:
Reproduced from the Wikimedia Commons drawing by Mariana Ruiz Villarreal, "File: Flagellum base diagram en.svg." In the public domain per http://en.wikipedia.org/wiki/File:Flagellum_base_diagram.svg.

pg. 230, FIGURE 13.2:
Climate Change 2007: The Physical Science Basis. Working Group I Contribution to the Fourth Assessment Report of the Intergovernmental Panel on Climate Change. Figure TS.6. Cambridge University Press. http://www.ipcc.ch/graphics/graphics/ar4-wg1/jpg/ts6.jpg.

pg. 233, FIGURE 13.3:

Reproduced from the Wikimedia Commons file "File: Carbon cycle-cute diagram.svg." http://en.wikipedia.org/wiki/File:Carbon_cycle-cute_diagram.svg. Original source: Lorentz, Katie. "Humans and the Global Carbon Cycle: A Faustian Bargain?" Langley Research Center. April 12, 2007. Image credit: NASA Earth Observatory. http://www.nasa.gov/centers/langley/news/researchernews/rn_carboncycle.html.

pg. 234, FIGURE 13.4:

Philippe Rekacewicz, UNEP/GRID-Arendal. "Temperature and CO2 concentration in the atmosphere over the past 400 000 years." *UNEP/GRID-Arendal Maps and Graphics Library.* http://maps.grida.no/go/graphic/temperature-and-co2-concentration-in-the-atmosphere-over-the-past-400-000-years.

INDEX